WORLD BANK TECHNICAL PAPER NUMBER 140

Environmental Assessment Sourcebook

Volume II
Sectoral Guidelines

AF680937

Environment Department

The World Bank
Washington, D.C.

Copyright © 1991
The International Bank for Reconstruction
and Development/THE WORLD BANK
1818 H Street, N.W.
Washington, D.C. 20433, U.S.A.

All rights reserved
Manufactured in the United States of America
First printing August 1991
Fifth printing June 1998

Technical Papers are published to communicate the results of the Bank's work to the development community with the least possible delay. The typescript of this paper therefore has not been prepared in accordance with the procedures appropriate to formal printed texts, and the World Bank accepts no responsibility for errors. Some sources cited in this paper may be informal documents that are not readily available.

The findings, interpretations, and conclusions expressed in this paper are entirely those of the author(s) and should not be attributed in any manner to the World Bank, to its affiliated organizations, or to members of its Board of Executive Directors or the countries they represent. The World Bank does not guarantee the accuracy of the data included in this publication and accepts no responsibility whatsoever for any consequence of their use. The boundaries, colors, denominations, and other information shown on any map in this volume do not imply on the part of the World Bank Group any judgment on the legal status of any territory or the endorsement or acceptance of such boundaries.

The material in this publication is copyrighted. Requests for permission to reproduce portions of it should be sent to the Office of the Publisher at the address shown in the copyright notice above. The World Bank encourages dissemination of its work and will normally give permission promptly and, when the reproduction is for noncommercial purposes, without asking a fee. Permission to copy portions for classroom use is granted through the Copyright Clearance Center, Inc., Suite 910, 222 Rosewood Drive, Danvers, Massachusetts 01923, U.S.A.

ISSN: 0253-7494

Library of Congress Cataloging-in-Publication Data
(Revised for volume 2)

International Bank for Reconstruction and Development.
Environment Dept.
Environmental assessment sourcebook.

(World Bank technical paper, 0253-7494 ; no. 139-140).
Includes bibliographical references.
Contents: v. 1. Policies, procedures, and cross-
sectoral issues — v. 2. Sectoral guidelines.
1. Economic development—Environmental aspects.
2. Environmental impact analysis. I. Title. II. Series.
TD195.E25I58 1991 333.73'14 91-4324
ISBN 0-8213-1843-8 (v. 1)
ISBN 0-8213-1844-6 (v. 2)

Volume II: Sectoral Guidelines

CONTENTS

FOREWORD

The Sourcebook is designed to assist all those involved in environmental assessment (EA). They include the environmental assessors themselves, project designers and World Bank task managers (TMs). This focus supports an important premise of EA, that sustainable development is achieved most efficiently when negative environmental impacts are identified and addressed at the earliest possible planning stage. The Sourcebook provides practical guidance for designing just such sustainable Bank-assisted projects.

Environmental assessment teams conducting EAs for borrowing governments need to know Bank policy regarding the project under consideration and which aspects of a project are of particular concern to the Bank. This Sourcebook provides specific information and common ground for discussion among those involved: EA professionals, the Bank and borrowing governments.

Project designers need to know applicable Bank requirements and the environmental implications of their design choices. In addition, they need to understand the objectives of an EA team. The Sourcebook provides not only project-specific considerations, but establishes common ground for general discussion, such as that regarding country strategy.

TMs are responsible for ensuring that borrowers fulfill Bank requirements for environmental review, including EAs where indicated. The Sourcebook provides assistance for these advisory tasks, through discussions of fundamental environmental considerations (with emphasis on those with relatively more impact); summaries of relevant Bank policies; and analyses of other topics that affect project implementation (e.g., financial intermediary lending, community involvement, economic evaluation).

Additional audiences likely to be interested in the Sourcebook are other economic development and finance agencies, EA teams for non-Bank projects, environmentalists, academics and NGOs.

The Sourcebook focuses on those operations with major potential for negative environmental impact, such as new infrastructure, dams and highways. Projects with relatively less negative potential, such as maintenance and rehabilitation, are not examined in detail at this stage; they merit a companion volume.

Bank policies and procedures, guidelines, precedents and "best practice" regarding the environment have been scattered throughout the institution and its publications -- or have resided only in the heads of Bank staff. This Sourcebook now collects this corporate knowledge into a single source. It is planned to be an easy-to-use reference manual, hence the overlaps and repetition. Its format is designed to facilitate the frequent updating necessary in a rapidly changing field such as the environment. The Table of Contents is the most efficient entry point from which to locate sections relevant for an individual user. Comments are invited at any stage from users on ways the Sourcebook can better meet their needs.

PREFACE
TO THE FIRST EDITION

The Environmental Assessment Sourcebook has been circulating for use as a draft for nearly one year. It is now offered to you for trial use. We seek your assistance on the present contents to alert us to any missing topics that should be included and we request further comments from "users." If at any time in the use of the Sourcebook you have comments, please let us know. The most efficient way to convey your views is to return marked up pages to my office (Room S-5029; or Fax: 202/477-0565).

The Sourcebook will be revised as new information and experience are acquired. The most up-to-date version will be available electronically for all people assessing the Bank's "All-in-one" electronic mail. With continued review and evaluation of experience, we expect to maintain this document as a living, working and up-to-date resource.

Kenneth Piddington
Director
Environment Department
The World Bank
1818 H Street, N. W.
Washington, D. C. 20433
U.S.A.

HOW TO USE THE SOURCEBOOK

The Sourcebook is designed to facilitate the environmental assessment process. It is intended to be used by all involved in EA, primarily the EA practitioner, but also groups managing them, project designers, task managers and environmentalists in general. While much of the document refers mainly to project loans, policy-based and adjustment lending may be addressed subsequently. The Sourcebook is a reference manual which contains the information needed to manage the process of environmental assessment according to the requirements of the World Bank's Operational Directive on EA (OD 4.00, Annex A, October 1989). It is a long document because of the wide range of subjects addressed. However, no one user will need all of the information in the book. Its contents have therefore been organized to be individually as easily accessible as possible, and there is a logical way in which a user can find the items that are pertinent to any particular lending operation. The Sourcebook focuses on operations with more potential for negative environmental impacts, such as major new infrastructure, rather than on operations with less potential impact, such as rehabilitation and maintenance, important though these investments undoubtedly are.

The Table of Contents is the most important section of the Sourcebook. It will assist the user of this reference manual who may be concerned about a specific operation (see the irrigation example below).

Chapter 1 is recommended reading for anyone responsible for a Bank-supported project with potentially significant environmental impacts. It summarizes Bank EA requirements and outlines the Bank's environmental review process, from screening at the time of project identification, right through to post-completion evaluation. A number of "boxes" illustrate different applications of EA in development activities. OD 4.00, Annex A is appended to Chapter 1, along with a list of other Bank operational policy and procedural documents relevant to EA. Annex 1-3 offers a standard format for Terms of Reference (TOR) for an EA that TMs may want to tailor to their specific needs.

Chapters 2 and 3 are "issues" chapters. They provide information and guidance on a number of topics, some of which are likely to arise in any EA. The issues in Chapter 2 are primarily ecological, while those in Chapter 3 are social and cultural. The chapters can, of course, be read in their entirety, but there are two other ways to use them. Their subtopics are shown in the Table of Contents, allowing the user to find them individually. They are also cited where applicable in the discussions of EA guidelines for specific project types, so that they can be referred to in the course of preparing to conduct a particular EA.

Chapters 4, 5, and 6 are "methods" chapters: economics, institutions, and financial intermediary lending. They are not intended to substitute for the knowledge and skills of experts carrying out the actual EA. Chapter 4 gives Sourcebook users an idea of what can be accomplished in the way of economic evaluation of environmental costs and benefits as part of an EA. Chapter 5 addresses institutional strengthening. It stresses the need to develop local capability in EA, identifies some of the broader needs for building country environmental management capacity that an EA may disclose, and considers what may be realistically expected in either area from a single loan or credit. Chapter 6 discusses the particular problems associated with EAs of sector and financial intermediary lending.

The extent to which these chapters are important to an individual user depends on the type of project and the nature of environmental management in the borrowing country.

Chapter 7, community involvement and the role of nongovernmental organizations in EA, explores the implications of OD 4.00, Annex A requirements in this area and offers guidance on how to meet them. Because community involvement is a new concept not only to some Bank staff but also to officials in borrowing countries, the chapter is recommended reading for all environmental assessors as well as task managers.

Chapters 8, 9 and 10 contain sectoral guidelines for EAs. The chapters begin with general considerations pertaining to EA in the sector(s) covered and with discussions of particularly relevant topics (e.g., "Integrated Pest Management and Use of Agrochemicals" in Chapter 8, which concerns the agricultural sector, or plant siting in Chapter 10, on industrial and energy sector projects). The topics are shown in the Table of Contents and cross-referenced throughout the Sourcebook. The balance of each chapter covers specific types of projects, chosen primarily because they have potentially significant environmental impacts. For each type, the project is briefly described (intended only to indicate the features of the project which have environmental significance), potential impacts are summmarized, and special issues are noted that should be considered in an EA. Possible alternatives to the project are outlined, and discussions of management and training needs and monitoring requirements are added. Each review concludes with a table of potential impacts and the measures which can be used to mitigate them. Sample Terms of Reference for the various project types are collected in one section in each chapter.

In the case of a loan for an irrigation project to reclaim arid land, the user would at a minimum consult the following Sourcebook sections:

Chapter 1: "The Environmental Review Process" (if not already acquainted with Bank EAs)

Chapter 8: "Irrigation and Drainage" (for the project-specific guidelines and sample TORs)

Chapter 2: "Arid and Semi-Arid Lands" and "Land and Water Resource Management" (for a review of ecological issues)

Chapter 7: "Community Involvement and the Role of NGOs in Environmental Review" (if not already familiar with the topic in Bank EAs)

The need for other information will become apparent; for example, tribal peoples, international waterways, new land settlement, resettlement, or institutional strenghtening may emerge as important concerns in the project, and the appropriate Sourcebook sections can be consulted.

ACKNOWLEDGMENTS

The Sourcebook staff is indebted to Bank colleagues and consultants who contributed to the first edition of The Environmental Assessment Sourcebook. We wish to express our gratitude to the members of the Environmental Assessment Steering Committee for overseeing this project from beginning to end. We thank our colleagues in the international community as well as colleagues in government and environmental agencies for their comments on various sections of the Sourcebook, and for sharing their own materials.

Environment Department: Kenneth Piddington, Director.

Environmental Assessment Sourcebook Staff: The Environmental Assessment Sourcebook was compiled and edited by Robert Goodland, Thomas E. Walton III, Valerie Edmundson and Charlotte Maxey.

Environmental Assessment Implementation Steering Committee: Gloria J. Davis, Chair, (ASTEN); Cynthia C. Cook (AFTEN); Colin Rees (ASTEN); Martyn J. Riddle (CENDD); J.A. Nicholas Wallis (EDIAR); Bernard Baratz, Stephen F. Lintner (EMTEN); Cesar A. Plaza (LATEN); Surinder P.S. Deol (POPTR); and James Listorti (Consultant).

Chapter 1: The Environmental Review Process: Author: Thomas E. Walton, III (Consultant). Reviewers: Cynthia C. Cook (AFTEN); Walter J. Ochs (AGRPS); Arthur E. Bruestle, Gloria J. Davis and Colin Rees (ASTEN); Thierry Baudon, Stephen F. Lintner, Spyros Margetis and Peter W. Whitford (EMTEN); Robert Goodland (ENVDR); and Albert Printz (Consultant).

Chapter 2: Global and Cross-Sectoral Issues in Environmental Review: Authors: Jan C. Post (ENVAP); Alcira I. Kreimer (ENVPR); Barbara Lausche (LEGOP); Barbara Braatz, Charlotte Maxey, Peter Little, Byron Nickerson, Richard Stoffle, Jon M. Trolldalen, James Talbot and Thomas E. Walton III (Consultants). Reviewers: Agnes Kiss, Walter J. Lusigi, Robert Tillman (AFTEN); David A.P. Butcher, Gloria J. Davis, Colin Rees and Susan S. Shen (ASTEN); Bernard Baratz, Stephen F. Lintner and Peter W. Whitford (EMTEN); Warren D. Fairchild (EMTAG); Robert Goodland (ENVDR); Alcira I. Kreimer (ENVPR); George Ledec (LATEN); Hans J. Peters (INUTD); Albert Printz and Lee Talbot (Consultants).

Chapter 3: Social and Cultural Issues in Environmental Review: Authors: David A.P. Butcher, Gloria J. Davis, Augusta Molnar and William Partridge (ASTEN); Mona Fikri, Jasper Ingersoll, Peter Little, Pam Stambury, Richard Stoffle and June Taboroff (Consultants). Reviewers: Michael M. Cernea, Scott E. Guggenheim (AGRPS); Poul A. Sihm (AFTAG); Cynthia C. Cook, Lee Talbot, Robert Tillman (AFTEN); Valter Angell, Arthur E. Bruestle, Colin Rees (ASTEN); Raymond J. Noronha (ENVAP); Mary B. Dyson and Michael P. Wells (ENVPR); John M. Courtney (ITFPS); Shelton F. Davis (LATEN); and Albert Printz (Consultant).

Chapter 4: Economic Evaluation Methods in Environmental Review: Authors: Herman Daly, Ernst Lutz and Mohan Munasinghe (ENVPR). Reviewers: Valter Angell (ASTEN); Jeremy J. Warford (ENVDR); John Dixon (LATEN); and Albert Printz (Consultant).

Chapter 5: Strengthening Local Capabilities and Institutions: Authors: Gloria J. Davis (ASTEN); Stephen F. Lintner (EMTEN); Barbara Lausche (LEGOP); and Thomas E. Walton III (Consultant). Reviewers: Jean B. Aden (ASTEN); Robert Goodland (ENVDR); Albert Printz (Consultant).

Chapter 6: Sector and Financial Intermediary Lending and Environmental Review: Authors: Martyn J. Riddle (CENDD); Charlotte Maxey and Thomas E. Walton III (Consultants). Reviewers: Rolf Glaeser (AF1IE); Paul Murgatroyd (AS1IE); Jean B. Aden (ASTEN); Kurt M. Constant (ASTIF); Paul A. Popiel (AFTTF); Robert D. Graffam and Rudolf van der Bijl (CCMDR); Millard F. Long (CECFP); Khosrow Zamani (CEMD2); Khalid Siraj (CODOP); Fred D. Levy, Jr. (EAS); Delbert A. Fitchett (EDIAR); Josef Duster (EM1AG); Bernard Baratz (EMTEN); Mark R. Nicholson (INVD1); Samia El Baroudy (LA1TF); and Christophe Bellinger (MIGGU); Melanie Johnson and Albert Printz (Consultants).

Chapter 7: Community Involvement and the Role of NGOs in Environmental Review: Authors: Robert Goodland (ENVDR) and William Nagle (Consultant). Reviewers: Francis J. Lethem (AF2DR); Cynthia C. Cook (AFTEN); Michael M. Cernea, Scott E. Guggenheim (AGRPS); Gloria J. Davis, David A.P. Butcher, William Partridge (ASTEN); Maritta Koch-Weser, Raymond J. Noronha, June Taboroff (ENVAP); Mary Dyson (ENVPR); David M. Beckmann (EXTIE); Shelton H. Davis (LATEN); Nancy Alexander (Friends Committee on National Legislation); Albert Printz (Consultant); and Diane Wood (World Wildlife Fund).

Chapter 8: Agriculture and Rural Development: Authors: Agnes Kiss (AFTEN); Robert Goodland (ENVDR); Anil Somani, Kirk Barker, Susan Braatz, Eugene Dudley, Peter Freeman, John Glenn, Charlotte Maxey, Byron Nickerson, James Talbot, and Thomas E. Walton III (Consultants). Reviewers: Poul A. Sihm (AFTAG); Cynthia C. Cook, Agnes Kiss, Robert Tillman (AFTEN); Guy J.M. LeMoigne (AGRDR); Shawki Barghouti, Walter J. Ochs (AGRPS); John F. Cunningham, Robert G. Grimshaw (ASTAG); Valter Angell, Roger S. Batstone, Susan Braatz, Arthur E. Bruestle, Colin Rees (ASTEN); Martyn J. Riddle (CENDD); J.A. Nicholas Wallis (EDIAR); Warren D. Fairchild, Permanand Gupta, Colin W. Holloway, Gert Van Santen (EMTAG); Bernard Baratz, Anders O. Halldin, Stephen F. Lintner, and Spyros Margetis (EMTEN); Robert J. Goodland (ENVDR); Asif Faiz (INUTD); Daniel Gross (LA1AG); William D. Beattie, Michael J. McGarry (LATAG); Dennis Child (USDA); Albert Printz, James Smyle (Consultants); and David J. Parrish (Virginia Polytechnic Institute and State University).

Chapter 9: Population, Health and Nutrition; Urban Development; Transportation; Water Supply and Sewerage: Authors: Carl R. Bartone (INURD); Albert M. Wright (INUWS); Sandra Cointreau, Colin Franklin, Peter Freeman, James Listorti, Charlotte Maxey, Byron Nickerson, James Talbot and Thomas E. Walton III (Consultants). Reviewers: Cynthia C. Cook, Robert Tillman (AFTEN); Shirin N. Velji (AS2IN); Roger J. Batstone, Arthur E. Bruestle, David G. Williams (ASTEN); Jean H. Doyen (AFTIN); Shirin N. Velji (AS2IN); Martyn J. Riddle (CENDD); A. Amir Al-Khafaji (EM2IN); Mario A. Zelaya (EM3IN); Maurice W. Dickerson (EM4IN); Stephen F. Lintner, Spyros Margetis, Peter W. Whitford (EMTEN); Richard A. MacEwen (EMTIN); Carl R. Bartone, Michael A. Cohen (INURD);

Asif Faiz, Jeffrey S. Gutman, Ian G. Heggie, Hans J. Peters (INUTD); Albert M. Wright (INUWS); John M. Courtney (ITFPS); Shelton H. Davis (LATEN); J. Rausche (United States Army Corps of Engineers); Perry Davies and Albert Printz (Consultants).

Chapter 10: Energy and Industry: Authors: Robert Goodland (ENVDR); Hans Adler, Sandra Cointreau, Eugene Dudley, Valerie Edmundson, Bernanda Flicstein, Ken Kosky, Tom Loomis, John Mulckhuyse, James Newman, Byron Nickerson, Anil Somani, James Talbot, and Thomas E. Walton III (Consultants). Reviewers: Robert Tillman, (AFTEN); John E. Strongman, Peter van der Veen (AFTIE); Uruj Ahmad S. Kirmani, Mihir Mitra, Christopher Wardell (ASTEG); Roger J. Batstone, Colin Rees (ASTEN); Martyn J. Riddle, Jean M.H. Tixhon (CENDD); David A. Craig (EM4IE); Suman Babbar (CFSPS); Bernard Baratz, Anders O. Halldin, and Stephen F. Lintner (EMTEN); Achilles Adamantiades, Mogens H. Fog (EMTIE); Anthony A. Churchill (IENDR); John Homer (IENGU); Alvaro J. Covarrubias, Hernan G. Garcia (LATIE); and Albert Printz (Consultant).

CHAPTER 8

AGRICULTURE AND RURAL DEVELOPMENT

Agriculture has been the Bank's largest lending sector for many years, and related environmental concerns are of proportionate significance. For task managers, this chapter of the Sourcebook outlines major environmental impacts of common types of projects while recognizing that every project has its own peculiarities and that the best procedure involves use of all resources available, particularly the expertise in appropriate technical departments.

As emphasized in the Sourcebook, the Bank views environmental assessment as an opportunity to identify early in the project cycle problems that become costly if they appear unexpectedly later. The guidance offered is a direction to proceed in. Not all environmental problems discussed in connection with one type of project will be found in the same project. And not all problems can be solved in one loan, especially when they transcend the limits of the project, as is the case with institutional weaknesses at sectoral or national levels. In agricultural projects, as in other sectors, task managers and governments will design more efficient and sustainable projects through the process of environmental assessment, and the procedures they develop will replace these early guidelines. In this way, increasingly flexible and cost effective measures will result -- to the benefit of everyone involved.

There are three classes of environmental concerns related to agricultural development. First are impacts from new land clearance or reclamation for any agricultural development. Second are impacts from intensification of production on existing agricultural land. Third are questions related to the sustainability of the agricultural developments.

LAND CLEARANCE OR RECLAMATION. Whenever new lands are brought into agricultural production for the first time, there are major initial impacts, and some of these are irreversible: loss of the natural resource cleared or reclaimed (e.g., forest, wetland, mangrove); soil erosion; declining soil productivity; extinction of species; loss of wildlife habitat; reduction in environmental services provided by existing ecosystems; and loss to forest dwellers or other vulnerable ethnic minorities who depend on the ecosystem that was removed. These impacts are dealt with in other Sourcebook sections, especially Chapters 2 and 3. In addition to these initial impacts, the method of clearing the land has great bearing on the subsequent success or failure of the project.

INTENSIFICATION. More conventional environmental problems with agriculture arise from intensification of production on existing agricultural land: soil compaction by heavy machinery, accelerated erosion, harmful impacts from overuse of fertilizers and biocides (see "Integrated Pest Management and Use of Agrochemicals" section), and soil salinization and improper drainage from irrigation schemes (see "Irrigation and Drainage" section).

SUSTAINABILITY. The crucial agricultural issue of sustainability, as mandated by OD 4.00 and by OMS 2.36, is amplified in Annex 1-5 of Chapter 1. Determining sustainability requires an evaluation of the economic, social and ecological costs of long-term continuation of the project; are they acceptable, and can they be sustained in relation to its outputs? Sustainability also is discussed throughout Chapter 8.

MANAGEMENT OF AGRICULTURE PRODUCTION

General Issues

1. In most circumstances, agricultural projects which are technically sound and sustainable will also be environmentally sound and sustainable. The exceptions to this principle are projects which result in the conversion of natural areas of ecological importance -- wildlands, mangroves, tidal marshes, tropical forests, etc. -- and projects in which indirect impacts such as induced development or uncontrolled hunting or logging in newly accessible wildlands are not prevented. In these latter cases, good agricultural practice cannot alone prevent environmental impact. Proper planning, environmental assessment, and mitigation will be necesary. For the task manager, then, the task of promoting compliance with the Bank's policies and objectives through EA involves:

(a) Seeing that the decision to convert a natural area to agricultural use is an environmentally acceptable one, and that the alternatives have been evaluated where there are environmental questions.

(b) Ascertaining that indirect impacts have been fully examined and taken into account in planning for project implementation and mitigation.

(c) Ensuring that good agricultural practices are followed in the design and implementation of projects in the sector.

2. It is not the purpose of the Sourcebook to define good agricultural practices. Information and guidance in that area is readily available from the Technical Departments. Instead, this chapter is intended to alert TMs to circumstances which may prevent application of good agricultural practice despite the intentions of the Bank and the project sponsors and to issues that may arise in environmental review of agricultural projects that fall into (a) or (b) above. Most of the questions to be considered are addressed elsewhere in the Sourcebook; cross-references are provided here.

3. Sustainable agriculture minimizes soil loss and maintains productivity through use of organic and inorganic inputs in balance with outputs. It takes into account land capability as a fundamental factor in any agricultural investment decisions. It recognizes that agricultural diversification is a key to the functioning of balanced upland farming systems and that external factors, such as road construction to improve market access may be critical in implementing diversification. (Sustainability is discussed in more detail in Annex 1-5 of Chapter 1.)

Agricultural Practices and Environmental Management

4. Soil conservation and cultivation practices intended to maintain productivity also minimize environmental damage from loss of vegetative cover, increased runoff, soil erosion and siltation. Judicious use of chemicals, both fertilizers and pesticides, is expedient for economic reasons and will either minimize or prevent eutrophication, groundwater contamination, nitrate accumulation, and evolution of pesticide resistance in non-target species which can result from excessive or indiscriminating

applications. Following guidelines for application rates and techniques is usually adequate to protect the environment, except where the substance is used is inappropriate for toxicological, biological reasons (see "Integrated Pest Management and Use of Agrochemicals" section). Recycling of manure and other wastes is common practice and is environmentally protective.

5. Good agricultural practice is likely to be used when it has been defined for the type of project, is managed at the project's scale of operation, advertised and explained to operators, and encouraged through inspection and incentives or sanctions. Practices should be proven to be practical, sensible, and profitable to the farmer, since he or she is ultimately the one who will decide if it should be used or not. This argues for strong links between the farmer and research. TMs should be alert to conditions that will prevent one or more of the above from occurring. Common impediments are listed below.

- national or sectoral policies that lead to short-term, minimum-cost exploitation of agricultural, forest, or fishery resources and penalize sustainable, higher-cost or slower pay-out operations;
- lack of land tenure or other sense of ownership and thus feeling of responsibility for the natural resource;
- practices being required for loans or credits involve structures or equipment beyond the means of the owners in small-scale operations;
- shift to intensive production techniques, such as animal feedlots, without accompanying technical assistance in areas such as waste management;
- lack of well-developed extension service or other outreach institutional structure for education and technical assistance;
- lack of attention to the differential access of male and female farmers to technologies and inputs;
- lack of environmental standards applicable to the sector, especially to intensive operations such as feedlots, fish farms, and processing plants;
- lack of capacity for field inspections to monitor adherence to standards and any required practices, such as for stream-bank buffer zones, reforestation, waste handling;
- lack of legal basis for enforcement action in cases of environmental damage or violation of requirements;
- lack of developed, alternative practices that are acceptable to farmers; and
- weak or non-existent provisions for dialogue between farmers and agricultural researchers.

6. Most of these questions fall within the "institutional" dimension of EA. They are critical to project success, because of the very fact that, unlike industrial or infrastructure loans or credits, where many of the significant impacts are avoided through design changes or "hardware", so many agricultural sector

operations proceed with the provision that there are potential impacts but that they will be avoided or mitigated merely by good practice. If the capacity to promote, implement and, where necessary, enforce good practice is flawed, adverse impacts can be anticipated.

Environmental Impacts

7. It is useful to divide agricultural projects into two catetgories: lowland, irrigated agriculture and upland, rainfed agriculture. Each has its own set of potential environmental problems. Impacts of agrochemicals and irrigation are primary concerns in the lowlands and are discussed in more detail in the sections on "Irrigation and Drainage" and "Integrated Pest Management and Use of Agrochemicals." In the uplands, problems of erosion, loss of soil fertility, improper land use and watershed management are more likely to be encountered. Sections on "Land and Water Resource Management," "Arid and Semi-Arid Lands" and "Forestry" are pertinent. Sustainability in agriculture will minimize soil and nutrient loss, and balance inputs with harvests, and strengthen the links between the farmer, extension and research. Environmental prudence, sustainability, and relevant technologies will be fostered by improving the links between farmer and research.

Major Impacts

8. The main environmental impact to be aware of is the irreversible loss of habitats. It is most critical when it is "wildland," but even degraded habitats, e.g., urban wetlands, perform valuable services. Such loss reduces economically valuable environmental services and accelerates extinctions and loss of wildlife. This loss can occur from two main causes: first, access roads that reach the project area; and second, clearing natural habitat for planting and processing of crops. Access roads leading into the project area or near habitat may facilitate unplanned settlements and destruction of that habitat. Loss of socio-economic services can result in increased pressure or conversion of other lands. If grasslands are converted, for example, graziers will have to graze their herds elsewhere. Indigenous peoples are particularly vulnerable. Similarly, when clearing new lands for the production or processing of crops, if the new land is wildland or other habitat, and especially if it is critical habitat such as tropical forests, (see "Wildlands" section), preventive measures, precautions and policies should be used for any guidance sought. The environmental and social impacts on the conversion of habitat types are discussed in Chapters 2 and 3.

Lesser Impacts

9. The lesser impacts are largely reversible and preventable. On occasion, however, they can be severe. The lesser impacts fall into only three categories:

(1) Agro-chemical runoff contaminating water and ground water ("Management of Agriculture Production: General Issues" section);

(2) Pesticide concerns ("Integrated Pest Management and Use of Agrochemicals" section); and

(3) Effluent disposal from crop processing ("Pulp, Paper and Timber Processing" and "Agroindustry" sections).

Prevention or mitigation of these impacts is relatively easy and should be standard procedure. They are discussed in the indicated sections, and in Goodland and others 1985.

Land Clearance Methods

10. The way in which new land is cleared for cultivation greatly influences subsequent agricultural success. For example, clearing tropical forest with heavy bulldozers or big chains decreases the thin layer of fertile topsoil, exposes infertile subsoil, and accelerates erosion. This issue is amplified in the sections on "Cross-Sectoral Issues" and "New Land Settlement" in Chapters 2 and 3.

Fuelwood

11. If fuelwood harvested from the wild is used in processing plants or for curing the crop, then that can be a major impact. (For further discussion, see "Natural Forest Management" section.) All fuelwood needed by such projects should come from fuelwood plantations planted near the processing plant as part of the project, and should not be harvested from the wild. The environmental assessment should ascertain if crop residues are used for fuel instead of having their nutrients and organic matter recycled to the crops.

Crop-Specific Impacts

12. Coffee, cotton, cocoa, oil palm, sugar and tea: apart from the major impacts from conversion of wildlands noted above, the environmental impacts accrue from processing -- mainly disposal of highly polluting washing waters -- as amplified in the "Agroindustry" section. The main point here is that most effluents can be profitably recycled. This is especially true for sugar factories, rubber and oil palm mill effluents which are valuable sources of the nutrients needed by the growing crop. Any crop residues, such as bagasse or cacao husks, not used for process steam also should be recycled to the extent possible as stock feed or cardboard, hardboard, or digested for fuel gas and fertilizer, etc., rather than being dumped in the nearest creek as was formerly common. Coffee and tea curing and sugar factories and other processing may consume fuelwood as mentioned in paragraph 10 above. Smoke from burning cane fields before harvest is an impact, but properly handled, it should cause more of a nuisance than an environmental problem. Cotton requires more pesticides than most other crops; this becomes a significant potential problem deserving special management.

Tobacco Production

13. Historically, the Bank has ceased to invest directly in tobacco production projects. Furthermore, Bank investments in tobacco, as components of rural development projects, have become almost negligible. Very small-scale tobacco may still be grown in mixed family farms as a minor crop in rural development supported by the Bank. These trends are expected to continue. The EA of projects containing tobacco production will identify the many impacts to such an extent that even tobacco components will decrease. The main considerations for the EA are:

(1) There is no safe level of use; tobacco products damage the health of the consumer and inquilines.

(2) Since tobacco is so profitable, the Bank, as lender of last resort, increasingly prefers not to use scarce resources for the purpose of competing with the private sector.

(3) The Bank can achieve much more development, specifically in the area of health, by encouraging a reduction in the consumption of tobacco products, by not investing in tobacco production, whether for domestic consumption or for export, than by promoting possible short-term producer profits.

(4) Much tobacco is cured using wood harvested from the wild. Such projects should routinely include sustainable woodlots. If the project cannot bear the costs of the needed woodlot, the EA will point out the major irreversible wildland conversion impacts which are not preventable in such cases, that the project itself will be in jeopardy (see "Wildlands" section).

On the other hand, the Bank is willing to invest substantially in promoting alternative crops for tobacco, in campaigns to reduce tobacco consumption, and in fire prevention.

Livestock

14. Although the Bank has no formal livestock policy, the main impacts are the same irreversibles noted above: access roads facilitating destruction of habitat, and the conversion of habitat for pasture. This is especially serious in the ranches created in moist tropical forest in Latin America, as well as being unsustainable in the short term. EA's for ranches to be created from tropical forest, will point to the major irreversible and unavoidable environmental costs.

15. Ranches in natural range routinely need constant monitoring of carrying capacity to prevent overgrazing in the dry years. Domestic cattle are so dependent on drinking liquid water every day that overgrazing is difficult to prevent around water sources. Therefore, the potential for using non-domestic livestock should be addressed during preparation. These do not need liquid water every day, do not suffer from tsetse and other insects as cattle do; hence, there is no need for biocides. Since they eat a much wider range of vegetation, as well as browse, they can produce more offtake than cattle over the medium and long term.

16. Partly because the Bank is concerned with poverty alleviation, it has not financed many livestock ranching projects recently, and this trend may continue. Grain can feed many more people than cattle can on the same area. The Bank's current emphasis is on more intensive peri-urban dairy projects, including stall-feeding, rather than the resource-wasteful extensive beef projects. (See "Livestock and Rangeland Management" section for more information on livestock projects.)

17. The lesser preventable and reversible impacts are the same as above: pesticide concerns, and processing effluents from slaughterhouses.

Fisheries

18. Freshwater fisheries normally have little environmental impact; on the contrary, they are usually highly benign, especially fishponds down slope from the household, which recycles household wastes and sewage leachate to great advantage. Highly concentrated aquaculture operations can pose a potential water pollution problem. In general, the Bank should do more to promote the fish potential in all relevant water projects, especially in reservoirs (see "Dams and Reservoirs" section).

19. The Bank has so few marine fishery projects that they are not a major problem. Sustainability should routinely be ensured by monitoring stocks and adjusting catch rates. Sea grass beds should be protected from damage by trawlers and draggers.

20. The major impact in shrimp projects is irreversible habitat conversion, especially of mangroves and other coastal wildlands and wetlands (see "Coastal Zone Management" section).

Social Impacts and Human Health

21. Involuntary resettlement occasioned by agricultural projects, especially irrigation reservoirs, is arguably their major impact. The complex issues surrounding involuntary resettlement have not historically been well handled in Bank-assisted projects; therefore, the Bank has issued detailed policies and guidance in this arena. This is especially so for indigenous peoples (see Chapter 3 for further discussion of "Involuntary Resettlement" and "Indigenous Peoples"). All agricultural projects should address how land is currently utilized within the farming systems, within the local economy, and as a source of nutrition, especially in poor crop years. When land is converted, the project should explicitly plan for substitutions or replacing these lost services, both for equity reasons and to avoid increasing pressure on other areas.

22. Fisheries, livestock development, and irrigation projects can have a negative impact on women's livelihoods and status when interventions increase women's burden without providing them additional assets or income. When mechanized fisheries activities (including shrimp) are introduced, women often lose labor opportunities because they lack new skills and access to marketing.

23. The main risks to human health are outlined for water-borne diseases in the section on "Dams and Reservoirs;" for risks from pesticide use, see the section on "Integrated Pest Management." Health risks resulting from processing cotton, such as byssinosis, are discussed in the "Public Health and Safety" section.

INTEGRATED PEST MANAGEMENT AND USE OF AGROCHEMICALS

1. Insects, weeds, pathogens and other pests1/ are a fact of agricultural life. They thrive on a concentrated and reliable food source and unfortunately, the measures commonly used to raise

---

1/ "Pests" refers to all animals, plants and micro-organisms which have a negative impact on agricultural production.

productivity of crops (e.g., monoculture with high-yielding varieties, multiple cropping with reduction or elimination of fallow periods, use of fertilizers, etc.) create an even more favorable environment for pests. Therefore, knowledgeable management of pest problems is required in any effective agro-system. The Bank's support of the integrated pest management (IPM) approach in its agricultural lending operations is perhaps best discussed in historical perspective.

2. The introduction of chemical pesticides2/ in the 1940s was widely regarded as a revolution in agriculture. They were relatively inexpensive and highly effective, and it became common practice to spray fields regularly throughout the season as a preventive measure even if there were no visible attack. Experience since then, however, has shown this approach not only to be environmentally damaging but also ineffective in the long term. Where pesticides have been used indiscriminately in this way, pest species have become resistent and difficult or impossible to control. In some cases, resistance in important disease vectors (e.g., malarial mosquitoes) has resulted or new agricultural pests have emerged. For example, all mite pests were created by pesticides since there were no mite pests before pesticides were used. Out of this experience, crop protection specialists devised a more diversified and sustainable approach, IPM, which is based on three fundamental principles:

(a) **To the extent possible, reliance is placed on using non-chemical measures to keep pest populations low.** For example, breeding and cultural practices are used to make the environment less hospitable to pests and to keep the crop healthy and resistant or tolerant to attack. This may include the introduction of non-indigenous pathogens or natural enemies.

(b) **The goal is to manage pests, not to eradicate them.** Populations of important pest species are monitored and control interventions are made only as necessary.3/

(c) **When pesticides4/ have to be used, they are selected and applied in such a way as to minimize adverse effects on beneficial organisms, humans and the environment.** OD 4.02 provides comprehensive guidelines for the selection and use of pesticides.

2/ The term pesticide is used here as it is better known than the more precise generic term biocide (literally: chemical killers of life). The term pesticide suggests that pests can be distinguished from non-pests, that pesticides will not kill non-pests, and that pests are wholly undesirable.

3/ The common statistical models used are the economic threshold level (ETL) and the action threshold level (ATL). The ETL is established through crop-loss assessment of the value of the crop, the amount of damage it can tolerate at each growth stage without significant effect on yield and market value, and the cost of crop-protection measures. The ATL is the pest population at which control action should be taken to prevent its reaching the ETL. Use of appropriate ATLs minimizes the frequency of pesticide application.

4/ "Pesticides" include insecticides, acaricides, herbicides, arboricides, rodenticides, fungicides, miticides, molluscicides, nematicides, etc., and also plant growth regulators (e.g., hormones and genetically engineered compounds).

3. Within these basic principles, approaches may vary depending on the crops or pests involved. The level of control necessary may also vary considerably, particularly for fruit and vegetable crops where cosmetic damage may significantly decrease market value.

Relationship to Bank Investments

4. Bank investments are to be designed according to the principles of environmentally sound and economically sustainable pest management as set forth in OD 4.02: "Agricultural Pest Management."5/ Pesticides may be one component of an agricultural pest management strategy in the context of an IPM program.

5. Such justifiable practice, however, requires special knowledge of the particular substances being used, and how they are stored and applied. It is important to remember that pesticides are, for the most part, toxic compounds that are dispersed in the environment and handled by large numbers of people. They vary greatly in degree of hazard to humans and the environment and in margin of safety for misuse.

6. Misuse can cause severe illness or death, contamination of soil and water, harm to livestock and wildlife, and even reduction or elimination of the natural enemies of the pests. Therefore, control and supervision of pesticide use will continue to be of critical concern in Bank projects.

7. The required level of evaluation of pesticide aspects will be determined by the task manager (TM) and the Regional Environment Division (RED) early in the project cycle. Factors that should trigger close attention during project preparation, appraisal and supervision, include: (a) significant financing or use of pesticides; (b) introduction or promotion of pesticide use in areas where they are currently little used; (c) introduction or expansion of crops (such as cotton, vegetables or rice) in which pesticide use is often high; (d) use of pesticide that is judged to present a significant hazard to health or the environment; or (e) questions concerning the likely return on investment in pest management.

8. For projects where none of these factors apply, it will usually be sufficient to pass these guidelines to the Borrower and provide assurance in the legal document that the guidelines will be followed. Bank policy clearly states that its requirements on pesticide selection and use are to be applied for any Bank-financed investment or adjustment operation under which pesticides may be procured, whether or not Bank funds are used for this purpose directly.

9. While pesticides are an easily recognized issue in agricultural projects where crop production is a goal, they are often overlooked where they are used in the reduction of post-harvest losses. Various types of pests (including molds) cause considerable losses in quality and quantity of stored foods worldwide and in developing countries the problem is often exacerbated by climatic conditions and lack of appropriate storage facilities. Produce in bulk storage is usually fumigated or treated with dilute insecticide dusts. Fumigation can be extremely hazardous to humans, requiring specialized equipment and training, and can leave potentially dangerous residues on the food.

5/ These draft guidelines deal only with agriculture (including forestry). Regarding control of disease vectors, see "Guidelines for the Use, Selection and Specification of Pesticides in Public Health Programs" (September 29, 198′), filed with OPN 11.01.

Bank Experience

10. Crop protection is an integral part of agricultural development and the Bank's approach has been to promote recognized "best practice" at the time. Consistent with this history, the Bank currently supports preparing a pest management plan for any agricultural development program, taking into account economic, environmental, and health and safety factors. Fundamental to this plan is an IPM program. In an IPM program, pesticides are used as a last resort. The Bank supports development and dissemination of IPM methods through research, extension and training, as well as through promoting supportive policy in the borrowing country.

11. The identification and preparation of any agricultural development project should cover anticipated pest problems and proposed pest management measures to determine: (a) the extent to which current practices are consistent with IPM; (b) problems and opportunities for improvement; and (c) the resultant strategy. The project and related operations (such as agricultural research or extension projects or sectoral loans) should be designed complementarily to fill gaps in knowledge, infrastructure and human resources, or to promote policy reform to facilitate the use of IPM.

Bank Policy, Procedures, and Guidelines

12. The Bank's policy on pesticides is stated in OPN 11.01: "Guidelines for the Selection and Use of Pesticides in Bank-Financed Projects and Their Procurement when Financed by the Bank" (March 1985). In preparation is OD 4.02: "Agricultural Pest Management." It provides guidance to staff on current Bank policies regarding use of the IPM approach in Bank agricultural lending operations and procedures for implementation of this policy. OD 4.02 is the guidelines for implementation of an IPM program, and it sets forth comprehensive guidance for the selection and use of pesticides.

13. Guidance on policies and procedures for procurement of pesticides will be provided in a Technical Note to be issued in conjunction with OD 11.01: "Procurement." The pesticide procurement guidelines address: (a) types of pesticides for which procurement should be prohibited or restricted; (b) specifications for product quality, packaging and labelling; (c) preparation of bidding documents; (d) qualification of bidders and after-sales service; and (e) evaluation of bids taking into account effectiveness, cost and human and environmental hazards.

14. Many pesticides are powerful and hazardous toxins which pose major risks (e.g., health damage to humans, creation of pest resistance, death of non-target species such as birds). Therefore, the EA should scrutinize the selection of pesticides with special care. Since it is impossible to be categorical, the Bank refrains from listing banned pesticides. However, the EA team and the TM should avoid broad spectrum persistent pesticides, and move towards IPM and the more modern less risky pesticides such as Bthuringiensis. All pesticides selected should be justified in the EA report.

Guidance for Environmental Assessments

15. IPM does not necessarily involve sophisticated information-gathering and decision-making procedures. It can be introduced at any level of agricultural development, for example, through variations in such basic crop management practices as planting times, crop spacing and residue disposal.

A useful beginning can be made with relatively little specialized information or management requirements. However, the effectiveness of any IPM program will be strengthened by the following:

(a) An understanding of the interactions between the elements of the local agro-ecosystem (e.g., crops, pests, beneficial organisms, the abiotic environment, etc.) and of any disruptions that may arise from overuse or misuse of pesticides.

(b) The development, with farmer involvement,6/ of a pest management plan using practical methods to reduce pest levels (including methods related to the total crop production system as well as those targeted specifically for pest control).

(c) The establishment of realistic economic and action threshold levels for key pests (see footnote 2).

(d) The development of practical systems for monitoring pest populations or infestation levels (either collectively or by individual farmers) and of whatever support structures are necessary to sustain them.

(e) The existence (or promotion) of farmer education concerning the principles and practices of IPM.

(f) The availability of appropriate materials and equipment, including establishment of insectaries to facilitate biological pest control where appropriate.

(g) Social and/or economic policy support that gives the farmer both incentive and opportunity to minimize pest management costs and increase productivity on a sustainable basis.

16. Similarly, research on crop protection/pest management methods does not necessarily have to be sophisticated, but it should be comprehensive. In addition to testing the efficacy (and environmental and toxicological characteristics) of pesticides, a complete research plan should include:

(a) Basic studies of the agro-ecosystem, including biology and life cycles of crops, pests and beneficial organisms and interactions among them (e.g., identification and classification of pests and natural enemies and evaluation of the impact of indigenous natural enemies on pest populations).

(b) Development and improvement of pest-resistant crop varieties.

(c) Examination of cultural practices for reducing pest populations (e.g., crop rotations, intercropping, timing of planting and irrigation, crop hygiene, land preparation, plant spacing, etc.).

6/ "Farmer involvement" means involvement of the people (often women) doing the work.

(d) Investigation of biological control methods, such as release of indigenous or introduced natural enemies, microbial pesticides, pheromones, repellents, etc.

(e) Identification of pesticides and techniques of pesticide use with minimal impact on beneficial organisms, humans and the environment.7/

(f) Determination of the actual impacts of different pests and population levels on crop yield and quality.

(g) Review of traditional pest management practices of local farmers, as these often have a sound biological basis.8/

17. Research carried out with the active participation of the intended beneficiaries is more likely to result in technologies that are practical under local field and socioeconomic conditions and thus to be adopted by the intended users. Therefore, an essential element of all pest management research supported by the Bank should be an integration of farmers and field workers into planning and implementation and an emphasis on trials in farmers' fields. The agricultural extension service should provide a useful link between farmers and researchers for these field trials.

18. Project support for IPM research can include: (a) technical assistance to evaluate and help develop pest management plans or research programs in specific areas; (b) training fellowships for researchers and technicians; (c) improving research facilities; and (d) grants for specific research projects and institutional strengthening. The Regional Environment Division can assist in identifying appropriate consultants or training institutions.

Training and Extension

19. The IPM approach should be the central focus of education and training related to crop protection at all levels: government policy makers, major buyers of crops, agricultural schools and colleges, pesticide suppliers, extension workers, farmers, etc. Education in IPM for farmers is particularly important because, in many countries, the farmers' primary alternative sources of information and advice

---

7/ For discussion of criteria to be used to determine which pesticides should be restricted (e.g., acute mammalian toxicity, environmental persistence, chronic health effects, toxicity to non-target organisms, etc.), see the World Health Organization's WHO Recommended Classification of Pesticides by Hazard.

8/ Sometimes, however, such practices are based on misperception of the importance of highly visible pests or may no longer be effective due to changing conditions. Also, natural compounds may have been developed that are extremely toxic. An attempt should be made to identify the substances and to control their use.

on such matters are the pesticide sales representatives or commercial middlemen who have little or no knowledge of the technical aspects of their safe use, and have strong incentive to emphasize benefits over risks.

20. The important role of natural enemies in controlling pest populations must be emphasized. Many farmers have no understanding of this concept nor of its underlying principles (e.g., many do not distinguish at all between beneficial and harmful insects). Not surprisingly, many farmers do not distinguish between the various pesticides (believing that any product is good for any pest in any crop), nor do they recognize the importance of using the correct dose at the right time. Because pesticides are often sold second- or thirdhand in remote rural areas, farmers may purchase products with no reliable information on what they are or how to use them. Extension workers should be prepared to advise farmers on the proper handling and use of pesticides, and on the hazards they pose to the farmers -- and to their families and livestock.

21. Training should therefore include a thorough understanding of pesticides: their effects and limitations, their associated health and environmental hazards, and requirements for safe and effective use and handling. In addition to conventional extension channels, a variety of methods and media should be used (e.g., radio, television, movies, illustrated pamphlets and comic books, etc.) to circumvent illiteracy and get the widest possible dispersion of information.

22. Special training, information and educational materials relating to IPM and pesticides should also be given to:

(a) Shopkeepers, vendors, farmers' groups or agricultural cooperatives involved in sale or distribution of pesticides, to ensure that all understand the toxic nature of the pesticides they are dispensing.

(b) Doctors, community health clinics and women's groups (particularly on symptoms and treatment of pesticide poisoning).

(c) Staff of credit institutions, to help them understand the economic advantages of IPM and to encourage them not to impose loan conditions that promote dependence on chemical pesticides or the use of very hazardous materials.

(d) School teachers and pupils (and their parents) in rural areas.

Policy Framework

23. While research, education and supply channels can provide the tools necessary for rational and cost-effective pest management, farmers must also have incentives to apply it. Therefore, in evaluating the crop protection component of a project, it is important to examine the economic policy framework and incentive structure and, if necessary, to propose measures to correct them. Clearly, when pesticide use is encouraged by subsidies from the government or donor agencies or by market requirements, the incentive for limiting their use is reduced.

Monitoring, Evaluation, and Supervision

24. Monitoring and evaluating progress on development and implementation of IPM programs is difficult because the scientific base may take a long time to develop, and also because farmers may not immediately adopt the technology offered. Often, therefore, progress must be measured initially just by correct processes being established: appropriate types of research and training programs (with adequate staffing and support) going on; strong linkages in place between research and extension; etc. Improvements relating to policy or infrastructure (e.g., the removal of pesticide subsidies), to safety of pesticide use and human exposure, and to environmental contamination (e.g., surface or ground water, wildlife) also should be tracked and evaluated. Again, market competitiveness is a major factor for the farmer and any successful IPM program will address that concern effectively.

25. When an IPM program is introduced in an area where pesticide use has been high, a reduction in pesticide application within a few years is probably a good indicator of successful implementation. A system of pest population monitoring should be established early in the project and linkage made between the results of such monitoring and control intervention decisions.

26. Where the necessary technical or institutional capability for such monitoring and supervision does not exist locally, provision to develop them should be planned. Formal or informal cooperation with local nongovernmental organizations (NGOs) should be sought. Where appropriate, Bank supervision should be readily available, including involvement by technical experts (local or international).

Contents of the Staff Appraisal Report (SAR)

27. The SAR for an agricultural loan should indicate how these guidelines have been followed in project preparation and how implementation will be evaluated and supervised. It should describe as fully as possible the anticipated pest problems and current pest management practices. Where these are not consistent with IPM principles, the SAR should include a schedule of the steps to be taken in the proposed projects to bring them in line. Areas of conflict should be discussed (e.g., malaria vs. swamp, drainage vs. use of insecticides). Pesticides selected should be explicitly justified in the SAR, together with training, monitoring, precautions, and other preventive measures to reduce environmental risks.

28. Similarly, the SAR should address current practices and problems, and propose measures for improvement regarding pesticide selection, use and disposal. Aspects to consider include:

(a) Existing or anticipated pest problems. Information on specific crop/pest targets should be provided to aid in identifying available IPM technologies that have been developed in other areas and may be adapted to local conditions. Information on pests that pose a public-health threat locally and control programs already in operation should also be provided, as these can influence agricultural pest management options.

(b) Proposed pest management plans and methods.

(c) Current institutional framework for crop protection (e.g., is pest management the responsibility of a government or company service or of individual farmers? are pesticides supplied by the government or through private commercial channels? are pesticides subsidized by the government? does market demand for a particular product quality encourage pesticide use?).

(d) The existing institutional structure and capacity for:

(i) dealing with pest problems (including diagnosis of pests and monitoring of population levels, impacts in the field and exchange of information on a regional basis);

(ii) applied research for development and testing, with farmer involvement, of pest management technologies (including biological control, intercropping, improved cultural practices and development of resistant varieties);

(iii) extension of pest management information and new technologies, and training of farmers; and

(iv) provision of services and inputs, and measures for cost-recovery for any services or inputs (including presence or absence of subsidies) provided by the public sector.

(e) Any aquatic or other important environmental resources in the proposed project area that may raise special concerns regarding pesticide use or other aspects of pest management (e.g., underlying aquifers used for domestic water supply, downstream commercial or natural fisheries or breeding grounds for aquatic species, adjacent natural parks or reserves, presence of endangered species that may be adversely affected).

(f) Legislation, regulations and enforcement capability relating to crop protection and to importation or use of pesticides. This includes: regulations concerning application procedures, training for application personnel, and application instructions accompanying products (fertilizers, fungicides, pesticides, etc.); and the capacity for reliable assessment of pesticide toxicity, human pesticide exposure and pesticide residues in foods and the environment.

(g) Technology packages extended to farmers for increasing agricultural production (usually including improved seeds, fertilizers, pesticides, cultural practices, etc.). IPM does not lend itself to standard packages; but in most cases, introduction of these packages has a significant impact on the pest situation, and vice-versa. Therefore, they must be evaluated with respect to their potential impact both on the agro-ecosystem and on prospects for adoption of an IPM approach.

AGROINDUSTRY

1. Agroindustry involves a diverse array of industries which process raw agricultural products. The main agroindustry subsectors include agriculture (and horticulture), forestry, fisheries and shellfisheries. This section reviews the following industries: cotton ginning, palm oil, tea and coffee, tanneries, slaughterhouses and wool scouring. Other agroindustries are reviewed under sections on "Food Processing," and "Pulp, Paper, and Timber Processing."

Potential Environmental Impacts

2. The major adverse impacts associated with agroindustry result from water and air pollution, disposal of solid wastes, and changes in land use. ("Solid Waste Collection and Disposal Systems" and "Wastewater Collection, Treatment, Reuse, and Disposal Systems" are discussed in more detail in Chapter 9.)

3. Wastewater streams vary with the type and size of the agroindustrial operation. Effluents typically have high biochemical oxygen demand (BOD), chemical oxygen demand (COD), and suspended and dissolved solids. Other contaminants such as pesticide residues, complex oils, alkaline or acidic compounds, and other organic constituents may also be present in wastewater. Discharges from feedlots, tanneries and slaughterhouses can be potential sources of disease among humans and animals.

4. Air emissions from agroindustrial operations commonly include particulate matter, sulphur dioxides, nitrous oxides, hydrocarbons, and other organic compounds. Noxious and nuisance odors are often associated with agroindustries.

5. Land resources may be adversely affected by inappropriate storage of raw materials and disposal of solid waste at the facility property or on off-site disposal areas.

6. The production of raw materials to supply these agroindustries can have negative environmental effects through intensified agricultural activity. Conversion of forest lands to agriculture has the potential for causing the most profound environmental and social impacts. The nature and degree of the impact will depend upon existing land use practices, the quantity of raw materials needed by the agroindustry, the production system chosen, and land and water management systems. Potential environmental impacts from intensified agricultural use include increased erosion, contamination of surface and groundwater from agricultural inputs (e.g., fertilizers, pesticides), changes in physical and chemical characteristics of the soil, and impacts on wildlife and native vegetation. Potential social impacts include decreased access to resources (e.g., traditional cropping and grazing areas, water resources, forest products), displacement of people, and social disruptions. (For further discussion of the potential negative environmental impacts asssociated with agroindustry projects and recommended measures to avoid or mitigate them, see Table 8.1 at the end of this section.) Not all the impacts, however, are necessarily negative. Planting of tree crops on degraded lands, for example, can have a positive environmental effect. The agroindustry can introduce more efficient agricultural practices into an area, stimulate markets for products and provide jobs for local people.

7. Indirect effects of the installation of a large agroindustry include the development of transportation facilities for delivering the products to markets and uncontrolled migration into the area of people looking for land or jobs.

8. The individual agroindustries and their main sources of pollution and potential health hazards are discussed in this section.

Palm Oil Industry

9. Palm oil is extracted from the outer pulpy portion of the oil palm fruit by a series of operations. The fruit is loosened in the bunches by steam sterilization. Strippers then separate the leaves and the empty bunches from the fruit. From the strippers, the fruit is conveyed to the digesters where it is heated and pulped. Free oil is drained from the digested pulp, then the pulp is squeezed and centrifuged to extract the remaining crude oil. The liquid is screened and clarified to produce purified oil. The extraction residues containing broken nuts and shells are further processed in a current of air to separate palm kernels from the shells. The palm kernels are dried, bagged, and stored for subsequent oil extraction, generally carried out at a different location.

10. Palm oil processing produces substantial quantities of solid wastes in the form of leaves, empty bunches, fibers, shells, and extraction residues. The bunch stalks contain significant quantities of recoverable nutrients and can create a major nuisance and disposal problem. Fibers, shells, and other solid residues are normally burned as fuel to produce steam. Uncontrolled burning of waste solids and release of air used to winnow the shell from the kernels contribute to air pollution.

11. Liquid wastes principally originate from sterilizers and oil clarification. The major pollutants are BOD, total suspended solids (TSS), COD, oil and grease (O/G), and organic nitrogen and ash.

Slaughterhouses

12. A simple slaughterhouse operation involves livestock holding pens, slaughtering, blood removal, hide removal or hog dehairing, evisceration, trimming and butchering for market. The main product of the operation is fresh meat as whole, half or quarter carcasses, or smaller meat cuts. The blood, hides, hair, and viscera are subject to further processing.

13. The main sources of pollutants are liquid effluents carrying varying amounts of solids. Odor from putrescible substances and organic decomposition is the only air pollutant but presents a continuous source of nuisance.

14. The most significant pollutants in the wastewater from slaughterhouses are: BOD (biochemical oxygen demand), TSS, O/G and fecal coliform bacteria. Solid wastes are normally screened and reprocessed or disposed of in a landfill.

15. The main safety hazards associated with slaughterhouses are cuts and abrasions from knives and cutting tools, falls on slippery floors, burns and scalds from hot water and steam, injuries from lifting, electrical shock from improper use of electric tools or defective electrical insulation. The main health hazard potential is from animal diseases, such as brucellosis, anthrax, acute and chronic respiratory syndrome, skin diseases, erysipelas, glanders, tularemia and Q-fever.

Tea and Coffee Production

16. Tea leaves are processed to yield either green or black tea. The manufacturing of black tea starts with the "withering" of the leaves either by natural drying or by hot air. Withering is followed by rolling to press out the juices and beating to break up the leaves. The broken leaves are sifted and fermented

to achieve the final product quality. The fermented tea is dried, graded and sorted for packaging. Green tea is prepared by heating the leaves either in hot pans or by steam. The leaves are rolled, fired and rerolled to achieve the desired quality.

17. Tea production may result in some gaseous wastes from the drying operation. These emissions are considered to be unimportant in comparison with the discharges from coal or oil fed boilers used to produce heat. Liquid wastes from cleanup operations are also considered to be insignificant.

18. Freshly picked coffee cherries are processed initially either by the dry method or the wet method. The dry process is used for Robusta coffee and for much of the Arabica coffee in Brazil and Ethiopia. The coffee cherry is sun-dried and then milled to remove in one process the outer skin, dried mucilage, the parchment and the silverskin layers. The milling process is undertaken in large installations. The waste products may be used for fuel, and also have on occasion been used for livestock feed.

19. On the other hand, the wet process, used to prepare higher quality Arabica coffee can seriously pollute. The ripe cherries are first washed to remove light fruit and rubbish, then pulped to remove the outer skin and some of the underlying mucilage. The freshly pulped beans are then fermented in tanks. This enzymatic process breaks down the remaining layers of mucilage into an effluent that can cause a serious pollution problem if discharged directly into streams or rivers. After a final wash the coffee, now called "parchment", is dried in the sun or artificially. The parchment coffee is then hulled to remove the parchment and silverskin layers to yield the internationally traded "clean" or "green" coffee beans.

20. The wet process requires a large volume of water and can cause serious pollution problems. Water economy can be achieved through recycling much of the flow, which also concentrates the enzyme content of the water used for pulping and this speeds up fermentation. Water used for final washing can be returned directly to rivers, but other effluent should pass through seepage pits.

21. Health hazards from the final stages of preparing tea and coffee for the market may include irritation of bronchial passages and lungs from exposure to fine tea or coffee dust, and asthma from hypersensitivity to tea fluff.

Tanneries

22. Tanning is the process of converting animal hides into leather. The hides are dehaired, tanned by reacting with tanning agents, dyed and finished to produce finished leather. The four processes involved in tanning hides are: beamhouse, tanhouse, retanning and finishing.

23. In the beamhouse, hides are processed by degreasing, fleshing and dehairing in order to prepare them for the tanning operation. Wastewaters contain dirt, salt, blood, manure, oil, and grease, flesh, hair, etc. Much of the solids is recovered and sold to rendering plants. The waste is characterized by high alkalinity, sulfide, nitrogen, BOD, COD, dissolved and suspended solids, and oil and grease.

24. The purpose of the tanning process is to produce durable material which is not subject to degradation by physical or biological mechanisms. Prior to tanning the hides are subjected to bating and pickling, both of which produce wastewaters high in acids and salts. Tanning is accomplished by leeching the hides with chrome, vegetable tannin, alum, metal salts or formaldehyde. Wastewaters from the operation are substantial. The spent chromium tanning solution is relatively low in BOD, COD and

TSS, but can contain significant concentrations of chromium, a toxic metal. On the other hand, vegetable tanning solution is high in both BOD and color.

25. Retan, color and fat liquor operations constitute the third major step in the tanning operation. The three operations are usually performed in one drum and involve addition of tanning solution (retan), dyes, and oils to replace natural oils of the hides (fat liquor). The process generates high-strength, low-volume discharges containing oil and color.

26. Finishing operations include drying, coating, staking, seeding, pasting and washing. The last two operation generate high-strength, low-volume wastewaters.

27. The tanning process generates significant airborne particulate matter and hydrogen sulfide discharges. Other gaseous emissions occur from ammonia stripping and utility boilers. The process also generates solid wastes in the form of fleshing, sanding dust, hide trimmings, sludges, greases, etc., which are normally recovered and sold to rendering plants. Hair is sold as a separate by-product.

28. The main health hazard, aside from accidents, is dermatitis from contact with chemicals and hides. Other health risks result from exposure to excessive dusts, toxic chemicals, noise and anthrax.

Wool Scouring

29. Wool scouring is a wet process used to remove natural and acquired impurities from the fibers. The process uses soap alkali or non-ionic detergents to scour the woollen fibers. As a final step the fibers are rinsed and dried.

30. Part of the water used in the process is recycled and the rest is discharged. The discharged effluent contains wool grease, urine, faeces, sweat, blood and other impurities which makes the wool scouring effluent one of the strongest industrial wastes in terms of BOD. The raw effluent is also high in oil and grease and may contain sulfur, phenols and pesticides.

Cotton Ginning

31. Modern cotton ginning is a continuous process spanning the unloading of raw cotton to baling of processed cotton fibers. Ginning of cotton produces large quantities of solid wastes in the form of cottonseeds (which may used in animal feed) and gin trash, and releases air pollutants in the form of cotton dust and lint. To minimize the source of pink bollworm in gin waste, movement and disposal of cottonseed and trash may be tightly regulated in certain countries. Where regulations permit, cottonseed is sent to oil mills for extraction of oil. Gin trash may be disposed of by composting, fumigation, sterilization and incineration. In some countries open burning is practiced which creates nuisance, air pollution and odor problems.

32. The main health related issue in cotton ginning involves cotton dust. Exposure to excessive levels of cotton dust leads to byssinosis, a serious respiratory disease. Excessive noise can also be a problem in this industry.

Special Issues

Environmental Effects of the Production of Raw Materials

33. As already discussed, conversion of large areas of natural vegetation to crop or livestock production, or the intensification of agriculture have potential negative environmental and social impacts. These factors must be considered in decisions on whether to finance a proposed project and on siting. Following are some examples relating to the agroindustries considered.

34. One of the most obvious and well-publicized examples of resource degradation arising from livestock production is deforestation in the Amazon basin; large areas of the moist tropical lowland forests have been converted to pasture for cattle raising. Large-scale clearing has long-term impacts because of the system's inability to recover from loss of nutrients and ecological disruption.

35. Planting of large plantations of oil palms also is linked with cutting of tropical forests, particularly in Southeast Asia. Conversion of the natural forest to another forest crop is less disruptive ecologically than conversion to an annual crop, yet still poses the risk of species loss and the problems arising from monocultures and agricultural production in general.

36. Large-scale cotton cultivation is becoming controversial in some areas because of its serious environmental impacts. The rapid growth of cotton production in parts of West Africa, for example, is linked with deforestation, erosion and declining soil fertility. Cotton often is grown in ecologically poorer areas which have higher susceptibility to degradation. Soil exhaustion was less evident in traditional systems in West Africa which provided for a long fallow period. Under more intensive systems the fallow period has been shortened or eliminated.

Facility Siting

37. Site location plays an important role in determining the environmental impact of wastewater discharges and solid waste disposal and the costs and methods used for pollution control. The capacity of a site to absorb waste will depend on the quality and quantity of waste and the ambient conditions. Many agroindustries, such as slaughterhouses and tanneries, which are heavy waste producers and polluters, should not be located in environmentally sensitive areas or at locations where their wastes (after proper treatment) cannot be assimilated without environmental degradation. Conversely, food producing agroindustries should not be sited in areas heavily affected by industrial discharges because of the risk of food contamination.

38. Facility siting is a highly complex and often time consuming process involving developers, special interest groups, politicians and local and national authorities. Developing countries are becoming more sophisticated in facility siting programs. Local regulations must be factored into the selection of an appropriate site for the proposed agroindustry.

Residue Utilization

39. The generation of residues and their potential use or sale depends on the raw materials, the production processes, output specifications, the cost of raw materials and products, the regulations

affecting product quality and use, and the constraints, if any, imposed on residue discharge or disposal. The residues may be liquid, gas or solid or a combination of the three. Using, recycling or commercializing residues have two advantages: increasing the productivity and thus the economic performance of the industry, and reducing wastes which are potential sources of pollution.

40. A review of options for residue utilization should be included during project preparation. The following commodities and their by-products are good candidates for such a review:

- palm oil residues for nutrient recovery and fuel
- palm kernels for oil and animal feed
- slaughterhouse by-products for rendering and animal feed additives
- cotton seeds for oil, animal feed additives and fuel
- animal oil for cosmetics
- animal by-products (including hair) and wastes

Existing Pollution Regulations

41. A number of countries and the World Bank have established effluent guidelines for many sectors of agroindustry. Contaminants which are regulated under U.S. Environmental Protection Agency (EPA) standards, for example, include pH, TSS, BOD, COD, and heavy metals. Standards for oil and grease, fecal coliform, and ammonia have also been established for slaughterhouses and seafood processing industries.

42. Regulations vary by country and industry type and are highly subjective. In countries having no regulations, World Bank guidelines should be followed; where regulations do exist, the more stringent standards should apply.

43. Air quality standards exist for particulates, sulfur dioxides, and some organic compounds in many developed countries; fewer ambient standards exist in developing countries.

Project Alternatives

44. The main areas where alternatives exist for decreasing the potential for negative environmental impacts are in facility siting and operation.

Siting

45. Selection of a site for an agroindustrial facility is dependent on a number of economic, ecologic and sociopolitical concerns. Regardless of the product manufactured or processed, an environmentally ideal site is one which satisfies the following criteria:

- availability of local land and water resources adequate to supply the required quality and quantity of raw materials without causing unacceptable environmental impact (e.g., clearing of primary forests, wetlands or critical wildlife habitat; and intensification of agriculture which will result in loss of soil fertility, increased erosion, etc.);

- sufficient land area to provide planned and expanded facilities for storage of raw materials, processing and waste disposal;

- minimal displacement of people and houses;

- minimal conflicts with higher valued land uses such as agriculture, especially in marginal land areas where prime agricultural land may be at a premium;

- proximity to receiving waters capable of handling effluent discharge without significant impact on the biophysical and aquatic environment;

- easy access to social and physical infrastructures such as skilled labor pool, support industries, transportation network, energy supply, raw materials and potential market areas;

- adequate distance from tourist or recreational areas, office buildings and housing complexes to minimize the impacts of odors, noise, and pollutants; and

- minimal construction and operational impact of the facilities on rare, threatened or endangered species, their habitats, or other sensitive ecosystems.

Operation of the Facility

46. Agroindustrial operations employ a variety of processes and equipment. The type of product processed and the size of the operation determines the type of equipment used, the quality and the quantity of the effluent or waste produced, and thereby the need for pollution control equipment. The various options for pollution control equipment should be examined in light of their effectiveness at minimizing pollution. The type of equipment used in pollution abatement cannot be specified for all possible agroindustries. Generally, pollution control measures employ the following processes:

(a) Water Pollution

- lagoons
- neutralization
- sedimentation
- filtration
- flocculation
- activated sludge treatment
- spray irrigation

(b) Air Pollution

- electrostatic precipitators and baghouses
- activated charcoal filtration
- scrubbing with sodium hypochlorite (for odor control)
- compost filters to reduce nuisance from odors

Management and Training

47. The need for management and training pervades all aspects of agroindustrial projects in developing countries. The technical capabilities of industry personnel and government officials responsible for monitoring pollution abatement performance frequently needs strengthening. The number of in-country consultants qualified to prepare detailed environmental assessments (EAs) is usually small and many of these are academics who lack project or industrial experience.

48. The issue of environmental management for development projects is complex. Many countries have just begun work on developing screening and review procedures for projects, compiling environmental databases, and matching and adapting to local priorities and requirements the multitude of environmental regulations and technical guidelines now on the market. Many central and provincial governments do not have trained staff, adequate budgets, or adequate natural resource databases to keep pace with the numbers of projects being offered by development agencies to host governments.

49. Use of an Environmental Management Plan during project implementation offers one way to establish ground rules for environmental protection and provide guidance to the host country managers. At a minimum, the plan should focus on mitigative measures and the means for their implementation. These measures should be incorporated into the work plans of the project office and other associated government agencies.

50. Training may be needed among the following groups.

- Government professional staff: in design and evaluation of EAs; collection, analysis and interpretation of pollution and related health data; plant safety and health procedures for key agroindustries; inspection and enforcement procedures.

- Industry employees: in environmental issues relevent to their operation; government regulations; evaluation of pollution data, treatment options and operational data; air and water pollution control options; operation and maintenance of the specific equipment used at their facility; plant safety, sanitation and health procedures.

- Local professionals not associated with government and industry: in providing adequate consulting services and/or independent review for EAs and pollution abatement.

51. All of the above people should be provided with some means of keeping abreast of developing technology so as to incorporate the best and most cost effective pollution abatement systems.

Monitoring

52. Air and liquid effluent and solid waste management must be monitored to determine if the project is adhering to the environmental standards and practices agreed upon. The following factors should be monitored for any given agroindustrial facility:

- Waste streams and gaseous and particulate emissions for appropriate parameters. Where a particular discharge is consistently above the established emissions limits or the standard for the industry, corrective action should be taken. These actions could involve process or equipment modification, upgrading, and housekeeping changes.

- Quality of receiving water downstream and air downwind of the plant.

- Effects of solid waste management practices on soils, groundwater and surface water resources.

- Implementation of health and safety plan by periodic site inspections to assure that training protocols and personal protective equipment such as dust masks are being employed in the work place. Standard industry practices should be used. Documentation and records should reflect periodic review and corrective actions taken.

53. An important factor in the abatement of pollution in agroindustrial development projects is the simultaneous strengthening of both in-plant and government monitoring capabilities. It should not be assumed that industrial facilities in the developing countries will develop and undertake air and water monitoring programs and install treatment technologies without effective government surveillance and legal and regulatory enforcement. Likewise, improved monitoring does not result in the reduction of wastewater discharges or air emissions unless the technical capability to comply with the effluent standards is also improved. To operate a successful monitoring program, it may be necessary to introduce sampling equipment and laboratory protocols (or the analytical laboratory) to the host country.

Table 8.1. Agroindustry

Potential Negative Impacts	Mitigating Measures
Direct: Site Selection	
1. Siting of plant or facility complex on/near sensitive habitats.	1. • Location of plant in rural area away from estuaries, wetlands, or other sensitive or ecologically important habitats, or in industrial estate to minimize or concentrate the stress on local environment and services. • Involvement of natural resource agencies in review of siting alternatives.
2. Siting of agroindustry along water courses leading to their eventual degradation.	2. • Site selection examining alternatives which minimize environmental effects and not preclude beneficial use of the water body using the following siting guidelines: • on a watercourse having a maximum dilution and waste absorbing capacity • in an area where wastewater can be reused with minimal treatment for agricultural or industrial purposes • within a municipality which is able to accept the plant wastes in their sewage treatment system
3. Siting of agroindustry so that air pollution problems are aggravated.	3. Location of plant at a high elevation above local topography, in an area not subject to air inversions, and where prevailing winds are away from populated areas.
Direct: Agricultural Practices	
4. Environmental deterioration (erosion, contamination of water and soil, loss of soil fertility, disruption of wildlife habitat, etc.) from intensification of agricultural land use.	4. Control of agricultural inputs and cropping/grazing practices to minimize environmental problems.

Table 8.1. Agroindustry (continued)

Potential Negative Impacts	Mitigating Measures
Direct: Plant Operation	
5. Aggravation of solid waste problems in the area.	5. • For facilities producing large volumes of waste, incorporation of the following guidelines in site selection: • plot size sufficient to provide a landfill or on-site disposal • proximity to a suitable disposal site • convenient for public/private contractors to collect and haul solid wastes for final disposal
6. • Water pollution from discharge of liquid effluents. • Plant: TSS; temperature; pH • Materials storage piles runoff: TSS; pH	6. Laboratory analysis of liquid effluent (including cooling water runoff from waste piles) in O/G, TDS, TSS, BOD, COD and in-situ temperature monitoring. All plants, or as indicated by agroindustry type • No cooling water discharge; if recycling not feasible, discharge cooling water only if receiving water temperature does not rise >3°C. • Maintain pH level of effluent discharge between 6.0 and 9.0. • Control effluent to EPA specified limitation (40 CFR 405-409;432) for specific process. • Land application of waste effluents where appropriate; the "Industrial Hazard Management" section should be consulted for guidelines regarding industrial hazardous materials.
7. Particulate emissions to the atmosphere from all plant operations.	7. Control of particulates by fabric filter collectors or electrostatic precipitators.

Table 8.1. Agroindustry (continued)

Potential Negative Impacts	Mitigating Measures
Direct: Plant Operation (continued)	
8. Gaseous and odor emissions to the atmosphere from processing operations.	8. Control by natural scrubbing action of alkaline materials; an analysis of raw materials during feasibility stage of project can determine levels of sulfur to properly design emission control equipment.
9. Accidental release of potentially hazardous solvents, acidic and alkaline materials.	9. Maintenance of storage and disposal areas to prevent accidental release; provide spill mitigation equipment.
Indirect	
10. • Occupational health effects on workers due to fugitive dust, materials handling, noise, or other process operations. • Accidents occur at higher than normal frequency because of level of knowledge and skill.	10. • Development of a Safety and Health Program in the facility designed to identify, evaluate, and control safety and health hazards at a specific level of detail to address the hazards to worker health and safety and procedures for employee protection, including any or all of the following: • site characterization and analysis • site control • training • medical surveillance • engineering controls, work practices and personal protective equipment • monitoring • information programs • handling raw and process materials • decontamination procedures • emergency response • illumination • regular safety meetings • sanitation at permanent and temporary facilities

Table 8.1. Agroindustry (continued)

Potential Negative Impacts	Mitigating Measures
Indirect (continued)	
11. Regional solid waste problem exacerbated by inadequate on-site storage.	11. Plan for adequate on-site disposal areas assuming screening for hazardous characteristics of the leachate is known.
12. Transit patterns disrupted, noise and congestion created, and pedestrian hazards aggravated by heavy trucks transporting raw materials and products to/from facility.	12. Site selection can mitigate some of these problems, but special transportation sector studies should be prepared during project feasibility to select best routes to reduce impacts; transporter regulation and development of emergency contingency plans to minimize risk of accidents.
13. Disease transmission from inadequate waste disposal.	13. Develop specifications for product preparation and/or processing, and waste disposal processes; monitor fecal coliform or other bacteria; and require documentation of waste disposal site monitoring.

Table 8.1. Agroindustry (continued)

Potential Negative Impacts	Mitigating Measures

The following guidelines are World Bank guidelines. If they cannot be achieved, the appraisal and/or supervisory mission should fully document the reasons for deviations. Where local regulations differ from those below, the stricter regulations should prevail.

Potential Negative Impacts

Leather Tanning and Finishing

Waste constituents:

Washing and soaking	BOD_5, TSS
Degreasing	BOD_5, TSS, O/G
Dehairing	BOD_5, TSS, pH, sulfides, nitrogen
Bating	Ammonia nitrogen
Pickling	Acids, salt
Tanning	Chromium, vegetable tannins
Retanning, coloring	Color, oil
Finishing	none

Mitigating Measures

Leather Tanning and Finishing

EPA Effluent Guidelines and Standards or use the following general guidelines.

Constituent	A	B	C	D	E	F
BOD_5	2.8	3.2	2.6	1.0	3.2	1.4
TSS	3.0	3.6	2.8	1.2	3.6	1.6
O/G	1.1	1.3	1.0	0.48	1.3	0.68
Sulfides	0.010	0.012	0.010	0.004	0.012	0.006
Tot. Cr.	0.10	0.12	0.10	0.04	0.12	0.06
TKN	0.54	0.64	0.50	0.20	0.62	0.28
pH	6.0 to 9.0					
Fecal Col.	Not over 400 per 100 ml					

Plant Categories by Primary Process

A: Beamhouse - pulp hair; tanning-chrome; finishing
B: Beamhouse - save hair; tanning-chrome; finishing
C: Beamhouse - save hair; tanning-vegetable; finishing
D: Beamhouse - hair previously removed; tanned previously; finishing
E: Beamhouse - hair previously removed or retained; tanning-chrome; finishing
F: Beamhouse - pulp or save; tanning-chrome or no tanning; no finishing

Table 8.1. Agroindustry (continued)

Potential Negative Impacts	Mitigating Measures
Palm Oil Mills Waste streams are acidic and have high TSS, O/G, and BOD_5.	**Palm Oil Mill Effluent Control** Limitations for liquid effluent include: pH 6 to 9 BOD <100 mg/l COD <1000 mg/l TSS <500 mg/l It is feasible to achieve 100 percent reduction of pollutant and waste effluent discharged to surface waters by one of the following: • spray irrigation • land application • evaporation ponds • discharge to municipal sewage treatment systems
Slaughterhouses Most important liquid effluent parameters are: BOD_5, TSS, O/G, pH, and fecal coliform organisms.	**Slaughterhouse Maximum Daily Effluent Limitations** (see table below) MPN fecal coliform count <400 per 100 ml

Plant Type	BOD (a)	TSS (b)	O/G	pH
simple	0.12	0.20	0.06	6-9
complex	0.21	0.25	0.08	6-9

Live wt killed: a) per kg; b) per megagram

Table 8.1. Agroindustry (continued)

Potential Negative Impacts	Mitigating Measures
Wool Scouring	
Raw waste contains significant quantities of O/G whose biodegradability constitutents a special problem.	Grease recovery step is necessary to decrease potential pollution from O/G.
Sulfur is brought in with wool, as well as phenolic and other organic compounds; this could be discharged to receiving waters.	Liquid Effluent Guidelines: Mg = megagram; mg = milligram BOD_5 5 Kg/Mg product TSS 4 Kg/Mg product COD 20 Kg/Mg product O/G 7.2 mg/l Total Chromium 0.1 mg/l Phenol 0.1 mg/l Sulfide 0.2 mg/l Pesticides 0.01 mg/l

DAMS AND RESERVOIRS

1. Projects for building dams and their associated reservoirs are usually planned for one or more of the following purposes: hydroelectric power, irrigation, domestic and industrial water supply, and flood control. ("Hydroelectric Projects" are discussed in Chapter 10; information on "Irrigation and Drainage" projects appears later in this chapter.)

Potential Environmental Impacts

2. Large dam projects cause irreversible environmental changes over a wide geographic area and thus have the potential for significant impacts. Criticism of such projects has grown in the last decade. Severe critics claim that because benefits from dams are outweighed by their social, environmental and economic costs, the construction of large dams is unjustifiable. Others contend that in some cases environmental and social costs can be avoided or reduced to an acceptable level by carefully assessing potential problems and implementing cost-effective corrective measures.

3. The area of influence of a dam project extends from the upper limits of the catchment of the reservoir to as far downstream as the estuary, coast and off-shore zone. It includes the watershed and river valley below the dam. While there are direct environmental impacts associated with the construction of the dam (e.g., dust, erosion, borrow and disposal problems), the greatest impacts result from the impoundment of water, flooding of land to form the reservoir and alteration of water flow downstream. These effects have direct impacts on soils, vegetation, wildlife and wildlands, fisheries, climate and especially the human populations in the area.

4. The dam's indirect effects, which on occasion may be worse than the direct effects, include those associated with the building, maintenance and functioning of the dam (e.g., access roads, construction camps, power transmission lines) and the development of agricultural, industrial or municipal activities made possible by the dam.

5. In addition to the direct and indirect effects of dam construction on the environment, the effects of the environment on the dam must be considered. (See Table 8.2 at the end of the "Dams and "Reservoirs" section for examples.) The major environmental factors affecting the functioning and life span of the dam are those caused by land, water and other resource use in the catchment above the reservoir (e.g., agriculture, settlement, forest clearing) which may result in increased siltation and changes in water quality in the reservoir and river downstream.

6. The benefits of a dam project are flood control and the provision of a more reliable and higher quality water supply for irrigation, domestic and industrial use. Dams may also provide an alternative to activities with potential for greater adverse impacts. Hydropower, for example, provides an alternative to coal-fired thermal power or to nuclear power. Intensification of agriculture locally through irrigation can reduce pressure on uncleared forest lands, intact wildlife habitat, and areas unsuitable for agriculture elsewhere. In addition, dams create a reservoir fishery and the possibilities for agricultural production on the reservoir drawdown area, which in some cases can more than compensate for losses in these sectors due to dam construction.

Special Issues

Hydrologic and Limnological Effects

7. Damming the river and creating a lake-like environment profoundly changes the hydrology and limnology of the river system. Dramatic changes occur in the timing of flow, quality, quantity and use of water, aquatic biota, and sedimentation in the river basin.

8. The decomposition of organic matter (e.g., trees) on the flooded lands enriches the nutrients in the reservoir. Fertilizers used upstream add to the nutrients accumulating and recycling in the reservoir. This not only supports reservoir fisheries, it also stimulates the growth of aquatic weeds, such as water lettuce and water hyacinth. Weeds and algal mats can be expensive nuisances when they clog dam outflows and irrigation canals, damage fisheries, curtail recreation, increase water treatment costs, impair navigation and substantially increase water loss through transpiration.

9. If the inundated land is heavily wooded and not sufficiently cleared prior to flooding, decomposition will deplete oxygen levels in the water. This affects aquatic life, and may result in large fish kills. Products of anaerobic decomposition include hydrogen sulfide, which corrodes dam turbines and is noxious to aquatic organisms, and methane, which is a greenhouse gas. The main gas produced, carbon dioxide, also exacerbates greenhouse risks.

10. Suspended particles carried by the river settle in the reservoir, limiting its storage capacity and lifetime and robbing downstream waters of sediment. Many agricultural areas on floodplains have always depended on nutrient-rich silt to sustain productivity. As sediment is no longer deposited on the floodplain downstream, the loss of nutrients must be compensated by fertilizer inputs to maintain agricultural productivity. The release of relatively sediment-free waters can result in the scouring of the downstream riverbeds. Sedimentation in the reservoir, however, provides higher quality water downstream for irrigation, industry and human consumption.

11. Additional effects of changes in the hydrology of the river basin include altered levels of the water table both above and below the reservoir and salinization problems which have direct ecological impacts and affect downstream water users.

Social Issues

12. The benefits of dams often accrue to urban dwellers, agricultural interests and others living some distance away from the dam, but less so or not at all to those who bear the heaviest environmental and social costs of the dam construction -- namely the inhabitants of area inundated by the reservoir, and those living on the floodplain. Reservoir filling results in the involuntary relocation of hundreds of thousands (in some projects), causing profound social readjustment not only for them but for those already living in the resettlement areas (see "Involuntary Resettlement" section). For those remaining in the river basin, access to water, land and biotic resources often is restricted. Artisanal riverine fisheries and traditional floodplain (recession) agriculture are disrupted by changes in stream flow and reduced silt deposition. Floodplains of many tropical rivers are vast areas of great importance for human and animal populations; when the floodplains shrink, land use must change or populations are forced to

move. Water-related diseases (e.g., malaria, schistosomiasis, onchocerciasis, encephalitis) often increase in reservoir and irrigation projects.

13. Social and environmental problems arise from the controlled and uncontrolled influx of other people into the area, such as construction workers, seasonal laborers for agriculture and other dam-induced activities, and rural people who take advantage of increased access to the area provided by roads, transmission lines or improved river navigation (see "Induced Development" and "New Land Settlement" sections in Chapter 3). The consequences are health problems, overburdened public services, competition for resources, social conflicts, and negative environmental impacts on the watershed, reservoir and downstream river basin.

Fisheries and Wildlife

14. As mentioned, riverine fisheries usually decline due to changes in river flow, deterioration of water quality, water temperature changes, loss of spawning grounds and barriers to fish migration. A reservoir fishery, sometimes more productive than the previous riverine fishery, however, is created.

15. In rivers with biologically productive estuaries, both marine and estuarine fish and shellfish suffer from changes in water flow and quality. Changes in freshwater flows and thus the salinity balance in an estuary will alter species distribution and breeding patterns of fish. Changes in nutrient levels and a decrease in the quality of the river water can also have profound impacts on the productivity of an estuary. These changes can also have major effects on marine species which feed or spend part of their life cycle in the estuary, or are influenced by water quality changes in the coastal areas.

16. The greatest impact on wildlife will come from loss of habitat resulting from reservoir filling and land use changes in the watershed. Migratory patterns of wildlife may be disrupted by the reservoir and associated developments. Poaching and eradication of species considered to be agricultural pests have a more selective effect. Aquatic fauna, including waterfowl, reptiles and amphibians can increase because of the reservoir.

Seismic Threat

17. Large reservoirs may alter tectonic activity. Though the probability that they will induce seismicity is difficult to predict, the full destructive potential of earthquakes, resulting in landslides, damage to dam infrastructure, and possible dam failure must be considered.

Watershed Management

18. Increased pressure on upland areas above the dam is a common phenomenon caused by the resettlement of people from the inundated areas and by the uncontrolled influx of newcomers into the watershed. On-site environmental deterioration as well as a decrease in water quality and increase in sedimentation rates in the reservoir result from clearing of forest land for agriculture, grazing pressures, use of agricultural chemicals, and tree cutting for timber or fuelwood. Similarly, land use in the watershed of the lower river basin affects the quality and quantity of water entering the river. It is therefore essential that dam projects be planned and managed in the context of overall river basin and regional

development plans, including both the upland catchment areas above the dam and floodplain, and watershed areas downstream.

Project Alternatives

19. A variety of alternatives, such as the following, exist for the design and management of dam projects:

- avoid or postpone the need for dam construction altogether by reducing demand for water or energy by conservation measures, efficiency improvements, fuel substitution, or restrictions on regional growth;
- avoid construction of a dam whose primary purpose is for irrigation by expanding and/or intensifying agriculture on the river's floodplain or outside the watershed;
- investigate possibilities for siting the project on an already dammed river by diversifying the functions of that dam;
- site the proposed dam on the river where it will minimize the negative and social impacts;
- adjust dam height, inundation area, dam design, and operational procedures to minimize negative environmental impacts; and
- install several small dams instead of a single large one.

Management and Training

20. Responsibility for management of the dam project should be vested in a river basin authority or other centralized entity, which will:

- collect baseline data
- build and manage the dam
- produce a water-use master plan with management strategies for regulation of the reservoir
- approve permit requests for major withdrawals and wastewater discharges
- coordinate disease vector control
- plan for municipal water supplies and water treatment facilities

Intersectoral cooperation should be ensured at both the policy and field levels with government ministries responsible for agriculture, fisheries, forestry, range and livestock, health, wildlife, tourism, municipal and industrial planning, and transportation.

21. Environmental and socioeconomic units should be established within the river basin authority and on the project site. The units should be staffed by professionals in physical (e.g., pollution control), biological (e.g., disease vector control), and social (e.g., involuntary resettlement) disciplines.

22. Guidance for the river basin authority can be provided by an advisory panel of international environmental specialists to advise on environmental aspects of the project and on training and management needs.

23. Training should be provided where needed to assure that the above mentioned disciplines are adequately represented.

Monitoring

24. Factors to be monitored should include:

- rainfall
- stored water volume in the reservoir
- annual volume of sediment transported into reservoir
- water quality at dam discharge and at various points along the river (such as, salinity, pH, temperature, electrical conductivity, turbidity, dissolved oxygen, suspended solids, phosphates, nitrates)
- hydrogen sulfide and methane generation behind dam
- limnological sampling of microflora, microfauna, aquatic weeds and benthic organisms
- fisheries assessment surveys (species, populations. etc.) in the river and reservoir
- wildlife (species, distribution, numbers)
- vegetation changes (cover, species composition, growth rates, biomass. etc.) in the upper watershed, reservoir drawdown zone, and downstream areas
- increases in erosion in the watershed
- impacts on wildlands, species or plant communities of special ecological significance
- public health and disease vectors
- in and out-migration of people to area
- changes in economic and social status of resettlement populations and people remaining in the river basin

Table 8.2. Dams and Reservoirs

Potential Negative Impacts	Mitigating Measures
Direct	
1. • Negative environmental effects of construction: • air and water pollution from construction and waste disposal • soil erosion • destruction of vegetation, sanitary and health problems from construction camps	1. • Measures to minimize impacts: • air and water pollution control • careful location of camps, buildings, borrow pits, quarries, spoil and disposal sites • precautions to minimize erosion • land reclamation
2. Dislocation of people living in inundation zone.	2. Relocation of people to suitable area, provision of compensation in kind for resources lost, provision of adequate health services, infrastructure, and employment opportunities.
3. Loss of land (agricultural, forest, range, wetlands) by inundation to form reservoir.	3. Siting of dam to decrease losses; decrease size of dam and reservoir; protect equal areas in region to offset losses.
4. Loss of historic, cultural or aesthetic features by inundation.	4. Siting of dam or decrease of reservoir size to avoid loss; salvage or protection of cultural properties.
5. Loss of wildlands and wildlife habitat.	5. Siting of dam or decrease of reservoir size to avoid/minimize loss; establishment of compensatory parks or reserved areas; animal rescue and relocation.
6. Proliferation of aquatic weeds in reservoir and downstream impairing dam discharge, irrigation systems, navigation and fisheries and increasing water loss through transpiration.	6. Clearance of woody vegetation from inundation zone prior to flooding (nutrient removal); provide weed control measures; harvest of weeds for compost, fodder or biogas; regulation of water discharge and manipulation of water levels to discourage weed growth.

Table 8.2. Dams and Reservoirs (continued)

Potential Negative Impacts	Mitigating Measures
Direct (continued)	
7. Deterioration of water quality in reservoir.	7. • Clearance of woody vegetation from inundation zone prior to flooding. • Control of land uses, wastewater discharges, and agricultural chemical use in watershed. • Limit retention time of water in reservoir. • Provision for multi-level releases to avoid discharge of anoxic water.
8. Sedimentation of reservoir and loss of storage capacity.	8. • Control of land use in watershed (especially prevention of conversion of forests to agriculture). • Reforestation and/or soil conservation activities in watersheds (limited affect). • Hydraulic removal of sediments (flushing, sluicing, release of density currents).
9. Formation of sediment deposits at reservoir entrance creating backwater effect and flooding and waterlogging upstream.	9. Sediment flushing, sluicing.
10. Scouring of riverbed below dam.	10. Design of trap efficiency and sediment release (e.g., sediment flushing, sluicing) to increase salt content of released water.
11. Decrease in floodplain (recession) agriculture.	11. Regulation of dam releases to partially replicate natural flooding regime.

Table 8.2. Dams and Reservoirs (continued)

Potential Negative Impacts	Mitigating Measures
Direct (continued)	
12. Salinization of floodplain lands.	12. Regulation of flow to minimize effect.
13. Salt water intrusion in estuary and upstream.	13. Maintenance of at least minimum flow to prevent intrusion.
14. Disruption of riverine fisheries due to changes in flow, blocking of fish migration, and changes in water quality and limnology.	14. Maintenance of at least minimum flow for fisheries; provision of fish ladders and other means of passage; provide protection of spawning grounds; aquaculture and development of reservoir fisheries in compensation.
15. Snagging of fishing nets in submerged vegetation in reservoir.	15. Selective clearance of vegetation before flooding.
16. Increase of water-related diseases.	16. • Design and operation of dam to decrease habitat for vector. • Vector control. • Disease prophylaxis and treatment.
17. Conflicting demands for water use.	17. Planning and management of dam in context of regional development plans; equitable allocations of water between large and small holders and between geographic regions of valley.
18. Social disruption and decrease in standard of living of resettled people.	18. Maintenance of standard of living by ensuring access to resources at least equalling those lost; provision of health and social services.

Table 8.2. Dams and Reservoirs (continued)

Potential Negative Impacts	Mitigating Measures
Direct (continued)	
19. Environmental degradation from increased pressure on land.	19. • Choice of resettlement site to avoid surpassing carrying capacity of the land. • Increase of productivity or improve management of land (agricultural, range, forestry improvements) to accommodate higher population.
20. Disruption/destruction of tribal/indigenous groups.	20. Avoid dislocation of unacculturated people; where not possible, relocate in area allowing them to retain lifestyle and customs.
21. Increase in humidity and fog locally, creating favorable habitat for insect disease vectors (mosquitos, tsetse).	21. Vector control.
Indirect	
22. Uncontrolled migration of people into the area, made possible by access roads and transmission lines.	22. Limitation of access, provision of rural development and health services to try to minimize impact.

Table 8.2. Dams and Reservoirs (continued)

Potential Negative Impacts	Mitigating Measures
Indirect (continued)	
23. Environmental problems arising from development made possible by dam (irrigated agriculture, industries, municipal growth).	23. Basin-wide integrated planning to avoid overuse, misuse, and conflicting uses of water and land resources.
External	
24. Poor land use practices in catchment areas above reservoir resulting in increased siltation and changes in water quality.	24. Land use planning efforts which include watershed areas above dam.

FISHERIES

1. The major types of fisheries fall under two categories: capture fisheries and culture fisheries (i.e., aquaculture and mariculture).

2. Capture fisheries harvest wild stock. The major methods are: (a) actively seizing fish or shellfish in a net (seine, trawl) or trap; (b) tangling fish in a net (gill, drift or trammel); and (c) catching them with hook and line (anglers, longliners). Capture fisheries operate in marine, fresh and brackish waters and range from large commercial concerns to small-scale artisanal fisheries. Marine fisheries include offshore and near-shore operations. Inland fisheries include those on rivers, lakes, reservoirs and estuarine areas.

3. Culture fisheries (marine, brackish and fresh water, ranching, etc.) involve management of resources to increase fishery production beyond that which is normally available from wild stock. Fish farming achieves higher concentrations of fish or shellfish by raising them in ponds; containing them in naturally productive areas by using cages, pens or nets; providing substrates for the attachment of nonmotile animals (e.g., oysters); or adding fish or shellfish to natural habitats (e.g., reef seeding, clam grow-out areas).

Potential Environmental Impacts

4. Since the Second World War, strong and sustained demand for fish for human consumption and livestock feed has put increasing pressure on fishery resources. The increase in world catch from 20 million tons in 1950 to over 90 million tons in 1990 has been made possible by technological innovations, deployment of distant-water fishing vessels, and exploitation of new oceanic stocks and shoaling pelagic species. The rate of increase of total world landings, however, has decreased since the 1960s, and the present biological potential for fish production has been estimated at 100 million tons. Total demand for fish for human consumption and for fish meal is increasing faster than the increase in supply, and is expected to exceed 100 million tons by the year 2000. Only limited increases can be expected in marine catches, which account for 80% of total fishery production, because the majority of commercial stocks appear to be fully or overexploited. Yields from freshwater capture fisheries, which supply 10% of total production, are unlikely to expand much, as they too have reached their biological limits in many areas. Fish farming has the greatest potential to increase fishery production, but it is associated with many environmental problems.

5. As demand is approaching production limits, many fishery resources are deteriorating. Overexploitation is depleting certain stocks, and other human activities are affecting fishery productivity and aquaculture potential in fresh, brackish and saltwater systems. These impacts affect traditional and commercial fishery operations as well as recreational and related water-based tourism. Pollution from industrial, urban and agricultural areas, land use in watersheds and water management affecting river flows and sediment loads, and coastal development all are having negative impacts on fisheries (see Table 8.3 at the end of this section for further discussion). The direct impacts on the environment by capture and aquaculture operations, as well as external environmental impacts on the fishery resources are examined in this section.

Capture Fisheries

6. The major direct negative environmental impact of capture fisheries is overexploitation. Overfishing not only degrades the target fish population by changing its population size and structure, but affects other species linked to it in the food chain. Non-target species are also inadvertently harmed or killed by the use of certain fishing equipment and practices that do not catch the desired species exclusively or that are harmful to habitats. Trawling is of special concern because dragging nets across the bottom can damage benthic communities. Accidental damage to coral reef by anchors and divers can be significant. Lost or discarded fishing nets, traps and other fishing gear entangle and subsequently kill fish needlessly (i.e., "ghost-fishing"). Although almost universally prohibited, explosives and poison are still used by some fishermen. Not only are many of the fish that are indiscriminately killed, wasted, but these practices may destroy habitats (e.g., coral reefs). Finally, the risk of oil and fuel pollution from accidental spills increases with intensified fishing activity.

7. Fisheries are subject to a wide variety of environmental impacts caused by human activities. Man's impact on the condition of the oceans, which still are in reasonably good condition, is limited. Substantial man-induced resource degradation is evident, however, in freshwater and coastal systems.

8. The effects of inland water resource management and land use are apparent both locally and downstream, often all the way to coastal ecosystems. Land use changes in the watershed, such as clearing of forests and increased agricultural activity, often will affect the quantity and quality of water entering surface waters which in turn have an impact on aquatic populations. The construction of dams and reservoirs, irrigation schemes, and flood control measures interrupt the pattern of seasonal flooding necessary to many fish for breeding and growth, change seasonal flow patterns, alter water quality, and disrupt or destroy fish habitats. (See the following sections for more information: "Dams and Reservoirs"; "Flood Protection" and "Irrigation and Drainage.") Losses to riverine and floodplain fisheries caused by dam construction, however, can be compensated, at least in part, by the reservoir fishery which is created.

9. Pollution of river, lake and marine waters by sewage, industrial effluent, acid rain and agricultural chemicals can reduce the survival rate of aquatic organisms, contaminate fish and shellfish, and create human health problems. Eutrophication from nutrient-rich inputs, such as fertilizer runoff, domestic detergents and untreated sewage effluent, can lead to mass fish mortality or gradual decline in fish populations, changes in species composition, and algal and phytoplankton blooms which foul nets and may be toxic to humans. Another source of pollution is non-biodegradable debris (e.g., plastic materials), which is increasing in quantity and becoming a serious hazard to fish which ingest the material or are entrapped by it.

10. Coastal ecosystems, including estuaries, mangrove swamps, seagrass beds, salt marshes and coral reefs are highly productive fishery habitats and play important protective roles against waves and high tides from the sea and flooding and sedimentation from the land. Many areas are being damaged or destroyed by the effects of accelerated settlement and economic development on the coastal fringe. Coastal development activities often affect runoff and cause silting and sedimentation of breeding grounds, coastal fishing areas and coral reefs. Dredging, land reclamation, drainage of wetlands and destruction of mangroves can directly or indirectly destroy important breeding and nursery grounds for fish. Oil pollution from offshore oil exploitation and naval traffic can foul nets, taint or kill fish, and spoil aquatic habitats.

Culture Fisheries

11. Aquaculture and mariculture projects, by manipulating natural systems, inherently have more significant possible environmental impacts than capture fisheries. Pond culture should receive attention.

12. The most obvious effect is the clearing of land and establishment of the ponds. This can be most destructive in coastal areas, such as in mangrove swamps and other wetlands which are particularly sensitive to disruption. The worth of these areas' production and protection functions is frequently undervalued and the importance of these areas to local economies underestimated. Extensive systems which involve large areas of ponds managed with minimal inputs are particularly destructive by virtue of the extent of land converted to ponds. Fish ponds established in inland areas are often established on flat, marginal agricultural lands of lower economic or ecological value. Nonetheless, ponds constructed on these lands may conflict with traditional uses of the land (e.g., seasonal grazing, livestock watering) which are of critical importance to local residents.

13. Fish ponds may have both positive and negative effects on local hydrologic conditions by altering water flow and affecting groundwater recharge. Ponds located in a natural stream channel, for example, can help reduce flooding in the immediate area, serve as a trap for sediments carried in runoff, and through seepage, raise local soil moisture. If located in an area subject to flooding, water diverted from the ponds by dikes could cause flooding elsewhere.

14. Water management in the areas affected by fish ponds is crucial, as fish ponds can reduce the water supplies available for competing demands, such as for irrigation, domestic or industrial use. Traditional drinking water supplies and washing areas can be disturbed when streams are diverted to aquaculture ponds. Local groundwater can be depleted by withdrawals for ponds. In general, the establishment of aquaculture ponds which draw on scarce ground or surface water supplies, particularly in arid areas, should be avoided except where fish production can be integrated with other water use (e.g., reusing pond water for irrigation, cage culture, in irrigation canals).

15. Pond drainage water can pollute nearby aquatic environments. The extent of pollution will depend on the quality of the pond water as well as characteristics of the receiving waters. The type and intensity of pond management -- frequency of water exchange, inputs of fertilizer and chemicals -- will determine the quality of water in the ponds. Pond water is almost always more nutrient-rich than surrounding waters, but will be more so if fertilizers and feeds have been added to the pond to increase fish productivity. Chemicals used in the ponds (i.e., for pond sterilization, weed, insect and disease control, water quality regulation, and control of undesirable fish) can also contaminate local waters. The quality of the receiving waters at the time of release from the ponds, and their dilution and dispersion capabilities will determine the effect of pond effluent on the nearby aquatic environment.

16. Ponds often are stocked with larvae and juveniles captured locally. This can deplete wild populations of fish, and erode capture fishery operations in the area.

17. Other potential negative impacts of fish pond aquaculture arise from the use of exotics: negative impacts on the wild native species from the spread of disease and parasites by the exotic species or from escape of pond fish to the wild. Selective breeding also has a potential long-term impact by reducing genetic diversity within fish populations.

18. Lastly, ponds can increase the incidence of human disease in an area by providing habitats for water-borne or water-related disease vectors such as snails (schistosomiasis) and mosquitos (malaria, dengue and other arboviruses).

19. Although fish farming operations that involve raising fish in nets or cages have few potential negative impacts, they can cause problems if practiced too intensively. The concentration of pens may increase to such an extent that navigation is hampered, water circulation restricted, and water quality decreased. Similarly, rafts or pilings installed for the cultivation of nonmotile animals can cause navigational hazards.

20. The external impacts on aquaculture are similar to those affecting capture fisheries. These include agricultural, industrial, municipal or transportation activities that decrease water quantity, degrade water quality, and increase sediment loads in supply waters. Drainage from irrigated fields or runoff from other agricultural areas containing fertilizers and pesticides will also affect fish pond water quality.

Processing and Transport

21. The indirect impacts of fishery projects result from the processing and transport of fish and shellfish (see "Food Processing" section). The effluent from fish processing plants is high in organic matter (offal and blood), oil and grease, bacteria, nitrogen and suspended solids. Discharge of the effluent into waters that cannot adequately dilute and disperse the waste can result in anaerobic conditions and fish kills; increased turbidity which affects corals, seagrasses and other organisms settling out of solids which smother bottom dwelling organisms; oil and grease that cause ecological and aesthetic problems; and contamination of fish and shellfish.

22. Fish processing, which often requires large volumes of water, can compete with other demands on the water supply.

23. The processing and transport of fish in a large fisheries project may require substantial infrastructure development, including roads, port and harbor facilities, and power and water supplies (for icing and refrigeration. etc.). Such developments have their own attendant impacts and are discussed in more detail in the following sections: "Rural Roads"; "Electric Power Transmission Systems"; "Roads and Highways" and "Port and Harbor Facilities."

Special Issues

Socioeconomic Issues

24. The development of all fisheries must be concerned as much with proper management of the fish resource as with improving the welfare of, or avoiding negative socioeconomic impacts on, fishermen, fish farmers, people involved in marketing and seafood consumers. While fisheries development can have beneficial effects on human nutrition in an area, the growth of commercial fisheries which export fish to external markets may reduce the quality or quantity of fish available for local consumption by competing with local fishermen and destroying or degrading aquatic habitats. Traditional lifestyles, patterns of resource use and subsistence economies may be disrupted or subverted by the introduction of cash economies in rural development fisheries schemes. Finally, human health risks are posed by

contaminated or spoiled seafood, or as can be the case with fish ponds, by creating a habitat for vectors of water-borne or water-related diseases.

25. Integrating both technical aspects and socioeconomic needs of fishing communities requires that the community actively participates in planning and execution of development activities. This improves the possibility that the fishery resource will be properly managed, as traditional methods often are found to be based on sustainable practices. It will also help assure that fishery resources are shared equitably between large and small-scale producers. Attention should be paid to ensuring that benefits accrue fairly to different social groups, and that middlemen do not erode fishermen's earnings. Also, to the extent possible, local labor should be given preference over imported labor.

Pond Fish Farming

26. Although in theory a promising enterprise, pond fish farming has had a high rate of failure. The most common causes of failure are poor siting and improper pond management. Poor siting can lead to problems with soil, water supply, drainage, and conflicting land use practices. The most important issue in management is flushing or exchange of pond water which must be done frequently enough to prevent the deterioration of water quality in the pond.

27. These factors are especially important for ponds located on coastal wetlands, whose previously waterlogged soils (if rich in pyrite and organic matter) can form acidic conditions when exposed to air or highly oxygenated water. Production decreases when the pond water becomes acidic or its quality deteriorates in other ways. In extensive areas of coastal wetlands (e.g., mangroves, tidal marshes) where the impression is that land is "unused" and therefore available, more land may be cleared for ponds when the existing ones become unproductive. The cycle begins again in a syndrome called "shifting aquaculture". Such pond operations not only are uneconomical, but economically unjustifiable, for the productivity of the ponds often proves to be lower than that of the natural ecosystem which has been lost.

28. Institutional factors also affect the success rate of pond fish farming projects. The private sector has been proven to be more successful than governments in this area. Where fish farming is judged to be economically feasible, efforts should be made to encourage private enterprises to invest.

Introduction of Exotics

29. The introduction of exotic species for fish farming or capture fisheries is a controversial practice. Not only have introductions or transplantations been less successful than anticipated, but they may have a net negative effect. Introducing exotics into new environments almost always poses the risk of competition with and predation on the indigenous species. Although exotics are introduced to increase fisheries productivity, they may be responsible for a net loss in fishery production by reducing populations of native species. In addition, exotics have the potential for introducing diseases and parasites into the local aquatic environment. In general, the introduction of exotics should not be advocated for capture fisheries, and should be done only with extreme care and only after precautions are in place in fish farming operations.

Project Alternatives

30. The main alternative to launching a new fisheries project is to improve the efficiency of existing fishery operations. In some places post-harvest losses due to spoilage are very high, particularly in traditional/artisanal and small-scale fisheries located in remote rural areas without infrastructure facilities for handling, processing, storage and marketing of the catch. In addition to reducing post-harvest losses, measures can be taken to salvage fish now discarded and intensify use of all fish by developing new products and expanding markets.

31. An alternative to pond aquaculture in coastal areas is to develop ways of utilizing the natural productivity of the intact ecosystems (e.g., mangroves, salt grass marshes) instead of converting them to pond production. Intensive aquaculture (e.g., oysters, mussels, clams, scallops) using the natural environment account for more than 50 percent of total production in volume (not in value). This type of approach may prove very valuable where pond aquaculture is a too sophisticated technology.

32. Within a given project there are a number of options for its design and implementation which can reduce ecological, social and economic problems. They are as follows:

(a) Culture Fisheries

- ***Procedures and techniques:*** native vs. exotic species; stocking from wild or hatchery organisms; extent of use of fertilizers, feeds and chemicals; harvest methods such as seining and pond draining; marketing methods; selling fresh vs. preserving; preservation methods such as icing, freezing, drying, salting, and smoking.

- ***Siting:*** location, design, size.

(b) Capture Fisheries

- ***Procedures and techniques:*** number and size of boats; fishing methods and equipment used; catch goals; fishing areas; marketing methods; preservation practices.

Management and Training

33. Most coastal governments have formulated fisheries policies and development plans, but achievement of successful fisheries development is difficult. This is due to the following factors:

- it is difficult to predict the size of various fish stocks targeted by capture fisheries due to natural fluxes in population size;

- sound management of the resource requires reliable statistics on fishery stocks and catches over time, involves expensive scientific programs, and requires effective enforcement of plans and regulations;

- there are many conflicting demands on freshwater and coastal aquatic environments which must be balanced by governments;

- land and resource practices outside of the jurisdiction of government fisheries agencies have profound effects on the fishery resources;

- coordination between governments and local communities is often weak, and where national government policies conflict with local views, needs and customs, the policies are difficult, if not impossible, to enforce;

- collaboration between governments in the management of stocks which cross national boundaries is difficult; and

- national governments are unprepared to take over the authority and complex responsibilities outlined by the Law of the Sea legislation, and other laws protecting individual species.

34. Multidisciplinary approaches essential for planning the optimal use of inland water, coastal or near-shore habitats, choice of technology, prevention of pollution, and reduction of conflicts with other sectors such as agriculture and industry, will require integrated planning. Coordination with other line agencies, local water pollution control agencies, and the appropriate agencies responsible for conservation of critical coastal ecosystems is essential. Intersectoral approaches should be explored at the earliest possible stage in project preparation to identify potential issues and conflicts so that mitigation strategies can be proposed. It may be impossible to satisfy the interests of all parties involved under a national fisheries law or government venture agreements. Negotiations must be conducted on a continuing basis. It is essential that an institutional environment be fostered in which competing interest groups may reach mutually acceptable and enforceable compromises.

35. Knowledge of existing law and socioeconomic norms which regulate individual and community property and use rights to fishing grounds within a project area are fundamental to project design. National fisheries laws and joint venture agreements for fisheries should contain provisions that would help achieve resource management objectives and protect the environment. Any or all of the following detailed provisions would best be determined at the time of project design: number and size of boats, authorized fishing areas, catch quotas, fishing methods, and catch limits. In some instances, new regulations may be needed to protect habitats or community rights of access to certain fishing grounds.

36. Appropriate project design for fisheries that factors sound management and development of the resource require a range of skills in biology, ecology, economics, law, and engineering. Many of these skills are lacking in developing countries, requiring sufficient institutional support, both at the local fishermen level and within the government agency responsible for the sector. Provision of skills through technical advisory services, technical and managerial training to develop capacities at all levels, and pilot or research components to determine appropriate management of the stocks or pond culture in question may be required.

Monitoring

37. Factors to be monitored include:

(c) **Capture Fisheries**

- water quality (including pollution and oil spills)
- fish stocks (population size and structure)
- fish landings
- conformance by fishermen to regulations on equipment use, fishing areas, catch, fishing seasons
- presence of any discarded materials causing "ghost-fishing"
- effects of land use or water management on water quality and fishery resources
- condition of non-fish species, especially indicator species (those most susceptible to changes in water quality)
- contamination of fish or shellfish or presence of conditions which could lead to contamination (e.g., red tide, oil spills)
- condition of coastal zone habitats (mangroves, sea grass beds, coral reefs)

(d) **Culture Fisheries**

- water quality in fish ponds or water bodies containing traps, nets or attachment substrates for nonmotile organisms
- water quality of fish pond effluent
- water quality and quantity of fish pond receiving waters
- hydrologic effects of fish ponds
- effect of aquaculture on local capture fisheries (population size and structure, health condition)
- presence of fish diseases or parasites
- contamination of fish or shellfish
- increase in water-borne or water-related disease vectors or human disease attributable to fish pond establishment

(e) **Processing**

- water quality of influent to and effluent from fish processing plants
- changes in commercial and non-commercial (especially indicator) species downstream of processing plants

Table 8.3. Fisheries

Potential Negative Impacts	Mitigating Measures
Direct Impacts: Capture Fisheries	
1. Overexploitation of fisheries stock and long-term degradation of the resource base.	1. • Fisheries management for optimum sustained yield: • restricted harvests (minimum size limits, catch quotas, seasonal closures) • gear restrictions (trawl bans, specified net mesh sizes) • closure of areas (permanent reserves, periodic closures) • limited entry systems (licensing, exclusive access) • prohibited practices (use of explosives, drift nets) • consideration of sustainable traditional fishery practices and incorporation to extent possible in modern fisheries management systems
2. Capture of non-target species and habitat damage through use of certain equipment and fishing practices.	2. • Limitation or prohibition of use of such equipment and fishing practices. • Testing and pilot scale use prior to large scale introduction of new technologies. • Expanded use of fish by development of new products and markets.
3. Pollution from oil and fuel spills and leakage, and from bilge flushing.	3. • Public education programs on proper fuel and oil handling and bilge waste disposal. • Provision of storage and handling facilities, bilge evacuation and disposal services. • See "Inland Navigation" and "Port and Harbor Facilities" sections in Chapter 9.

Table 8.3. Fisheries (continued)

Potential Negative Impacts	Mitigating Measures
Direct Impacts: Capture Fisheries (continued)	
4. Diver and anchor damage.	4. • Public education programs for fishermen on effects of damage and ways to avoid it. • Installation of mooring buoys. • Designation of anchoring locations.
5. "Ghost-fishing" and navigational hazards from abandoned or lost fishing nets and traps.	5. Public education programs for fishermen on hazards of abandoning gear.
6. Use of explosives and poison.	6. Prohibition of practices and enforcement of regulations.
7. Introduction of exotics leading to degradation of native stocks.	7. Prohibition of exotics introduction.
Direct: Culture Fisheries	
8. Clearing/conversion of coastal wetlands for construction of ponds.	8. • Prohibition of ponds in area of particular ecological significance. • Limitation of area converted to ponds. • Intensified management in existing and new ponds to discourage "shifting aquaculture" and low-input, extensive aquaculture which converts large areas.

Table 8.3. Fisheries (continued)

Potential Negative Impacts	Mitigating Measures
Direct: Culture Fisheries (continued)	
9. Erosion and siltation problems arising in construction phase.	9. • Restriction of clearance to area needed for ponds. • Pond construction during dry seasons. • Stabilization of exposed soil with grasses or other ground covering.
10. Competition of ponds for water and land resources demanded by other users.	10. • Assessment of existing traditional land and water use and agricultural, industrial and municipal demands. • Planning, management and continuing negotiations to reach acceptable distribution of resources. • Siting of pond to avoid disruption of traditional uses of water for washing and drinking. • Coordination of aquaculture ponds with other activities to double-up on water use (e.g., pond water reused for irrigation).
11. Loss of productivity or formation of toxic conditions in ponds from high temperatures, low oxygen and waste accumulation.	11. Adequate pond water exchange and frequent pond flushing.
12. Acidification of pond water due to hydrogen sulfide formation.	12. • Siting in areas not susceptible to acidification (avoid waterlogged soils high in pyrite and organic matter). • Adequate pond water exchange and flushing.

Table 8.3. Fisheries (continued)

Potential Negative Impacts	Mitigating Measures
Direct: Culture Fisheries (continued)	
13. Local depletion of larval and juvenile organisms for pond stocking.	13. Production of larvae and juveniles in nursery.
14. Water pollution from pond effluent (nutrient-rich and with varying chemical content depending on intensity of pond management).	14. • Release into water body with adequate dilution and dispersal capability. • Dilution prior to release. • Timing of release with period of high water. • Shorter retention time of water in pond: more frequent pond water exchange and flushing. • Treatment of water prior to release.
15. Introduction of exotics with subsequent damage to native stocks by competition, predation, spread of disease and parasites.	15. • Avoidance of exotic introductions except where adequate knowledge of biology and life history of species indicates low risk of negative impacts and where adequate safeguards against escape are taken. • Regular monitoring for disease and parasites; if present and spreading, elimination of infected populations. • Consideration of using sterile hybrids.

Table 8.3. Fisheries (continued)

Potential Negative Impacts	Mitigating Measures
Direct: Culture Fisheries (continued)	
16. Spread of disease in aquaculture stocks and into natural stocks when stocking becomes too dense.	16. • Monitoring of disease incidence. • Limitation of numbers when disease is positively correlated with fish densities. • If disease spreads, elimination of diseased individuals.
17. Concentration of pens, pilings, and rafts in natural (non-pond) water bodies to extent that navigation is hampered, water circulation is restricted, water quality is decreased, and capture fisheries precluded.	17. Regulation of aquaculture activity to limit it to acceptable intensity.
External Impacts: Capture Fisheries and Culture Fisheries	
18. Dams which alter water quality and stream flows and disrupt riverine and floodplain fisheries.	18. • Establishment of reservoir fishery. Water release management to minimize damage to fisheries (see "Dams and Reservoirs" section).
19. Irrigation schemes which alter water quality and quantity.	19. • Development of fishery activities in conjunction with irrigation systems (e.g., use of pond water for irrigation, traps and nets in irrigation canals). • Management of irrigation schemes to minimize damage to fisheries (see "Irrigation and Drainage" section).

Table 8.3. Fisheries (continued)

Potential Negative Impacts	Mitigating Measures
External Impacts: Capture Fisheries and Culture Fisheries (continued)	
20. Land use and agricultural practices in watersheds affecting sediment content and water quality.	20. • Integrated watershed planning and management. • Close coordination between fisheries and government agencies responsible for resource management in watersheds to alert them to impacts on fisheries.
21. Flood control measures damaging to water quantity and quality and aquatic habitats.	21. See "Flood Protection" section.
22. Pollution from industrial effluent, sewage and agrochemicals affecting fish survival and tainting fish.	22. See "Wastewater Collection, Treatment, Reuse, and Disposal Systems" section in Chapter 9; see also the sections on "Plant Siting and Industrial Estate Management" and "Industrial Hazard Management" in Chapter 10.
23. Air pollution and acid rain affecting fish survival.	23. See "Atmospheric Pollution" section in Chapter 2.
24. Coastal development involving dredging, filling, destruction of mangrove swamps, construction and infrastructure development.	24. See "Coastal Zone Management" section; "Port and Harbor Facilities" projects are discussed in Chapter 9.
25. Oil pollution from coastal and inland navigation, and spills from drilling, transport operations, and oil tankers (cleaning of tanks).	25. See the following sections in Chapter 9: "Inland Navigation" and "Port and Harbor Facilities"; "Oil and Gas Development--Offshore" and "Oil and Gas Development--Onshore" are examined in Chapter 10.
26. Water-based tourism development which conflicts with fishing activities.	26. See "Tourism Development" section in Chapter 9.

Table 8.3. Fisheries (continued)

Potential Negative Impacts	Mitigating Measures
Indirect Impacts: Capture Fisheries and Culture Fisheries	
27. Creation or expansion of port areas, shore facilities and infrastructure (roads, water, power) for processing and transport of fish products.	27. See "Plant Siting and Industrial Estate Management"; "Coastal Zone Management"; and "Port and Harbor Facilities."
28. Pollution from effluent discharged from industrial fish processing plants.	28. • Discharge into waters with adequate dilution and dispersal capabilities. • Water quality monitoring for suspended solids, oil and grease, dissolved oxygen, nitrogen and coliform. • Reduction of wastes by recycling into usable products, reduction of water use. • Treatment of waste prior to release.

FLOOD PROTECTION

1. Flood protection includes both structural and non-structural means to provide protection from floods or reduce the risks of flooding. Structural flood control measures include dams and reservoirs, river channel modifications, dikes and levees, overflow basins, floodways and drainage works. Non-structural measures include regulation of floodplain uses through zoning, floodplain regulations, building and sanitary ordinances, and regulation of land use in watershed areas.

Potential Environmental Impacts

2. Except for unusually severe flooding, ecosystems, and in many areas, human communities, are adapted to and rely on periodic inundation of the land. Flooding usually becomes a problem only when natural events or human activities increase flooding intensity or frequency, or man invades flood-prone areas with structures and developments that need to be protected.

3. The major potential environmental impacts of structural flood control measures arise from the elimination of the natural pattern of flooding and the benefits flooding confers. Floodplains are productive environments because flooding makes them so; flooding recharges soil moisture and replenishes the rich alluvial soils with flood deposits of silt. In arid areas flooding may be the only source of natural irrigation and soil enrichment. Reduction or elimination of flooding has the potential for impoverishing floodplain (recession) agriculture, natural vegetation, wildlife and livestock populations on the floodplain, and floodplain and riverine fisheries which are adapted to the natural flood cycles. To maintain the productivity level of the natural systems, compensatory measures have to be taken, such as fertilization or irrigation of agricultural lands, rangeland improvement, or intensified fisheries management and production systems. In addition, when channelization measures reduce the frequency of flooding, the sediments entering the river systems from watershed areas upstream will be passed to the mouth of the river unless overflow areas are present downstream. Increased sediment loads in the river can cause physical changes through sedimentation and altered water flows in the estuary, delta or near-shore coastal areas, and affect rich fisheries supported by these ecosystems. Coral reefs, particularly sensitive to increased sediment outflow from rivers, can be irreparably damaged. (A list of flood control measures is provided in Table 8.4 at the end of this section.)

4. The function of dams and reservoirs in flood control is to reduce the peak flows entering a flood prone area. Reservoir operation for flood control is distinctly different than that for hydroelectric or irrigation use. Rather than maintaining high water levels for increased head or sustained water supply for irrigation, flood control operation requires that water levels be kept drawn down deliberately prior to and during the flood season in order to maintain the capacity to store any incoming floodwater. The eventual release of water, however, may pose problems. Instead of being flooded to a greater depth for a shorter period of time, some lands may be inundated to a much shallower depth but for a considerably longer period. This may not be compatible with existing agricultural systems. The environmental effects of dams and reservoirs, covered in detail in the preceding section, will not be discussed further here.

5. Structural flood control measures such as levees and channel improvements increase the capacity of a stream by increasing the size of the channel, increasing the velocity of flow, or achieving both

simultaneously. Modification of the channel includes dredging to make it wider or deeper, clearing it of vegetation and other debris, smoothing the channel bed and walls, or straightening the channel, all of which help increase the rate at which water is passed through the system, thus preventing flooding. Straightening the channel by eliminating meanders also helps reduce the risk that water will breach the river bank on the outside of curves where the current is most rapid and water rises highest.

6. Channel modification can provoke a number of negative environmental impacts. Any measure that increases the velocity of flow increases the erosive capacity of the water. Problems of erosion and sedimentation can arise both on-site and downstream. Paving the channels reduces or eliminates all factors that retard flow, but the practice poses many aesthetic and ecological problems, including a reduction of groundwater recharge and disruption of aquatic populations. Channel smoothing and clearing as well as dredging also have a great impact on aquatic organisms and fisheries by disrupting their habitats. Disposal of dredging spoils can create another set of problems. Although channel improvement can alleviate flooding problems in the treatment area, flood peaks are likely to increase downstream, thus simply transferring the problem elsewhere.

7. Artificial levees, improved natural levees and dikes increase channel capacity and contain all but exceptionally high floodwater in the channel. As with channel improvement measures, however, these structures tend to pass floodwater on to downstream areas which in turn suffer or are forced to spend public funds to implement flood control measures themselves. Dikes built on the floodplain to exclude water from certain areas affect the hydrology of the area, and can have impacts on wildlife and livestock habitat and movement.

8. Overflow basins usually are swampy areas between river levees and valley sides. Artificial basins, into which floodwater is diverted, can also be constructed. On-site detention basins, or small impoundments, used most often near urban areas to intercept and collect runoff before it reaches the stream channel, are also effective in reducing peak flows. The basins have a positive impact in recharging groundwater aquifers and settling suspended sediments which would flow into the channel. They can, however, provide habitats for disease vectors.

9. Floodways (also called high-flow diversions or spillways) are natural or man-made bypass channels or conduits that redirect waters around or away from urban centers or areas of high population density. Further downstream the water can be re-diverted into the river from which it originated.

10. Flood-control structures are costly to build. They also give a false sense of security because people think that the risk of flooding has been eliminated rather than diminished. This may encourage them to increase development on the floodplain with disastrous results in the event of an unusually high flood or if control structures fail.

11. In addition to the impacts of flood control structures on the environment, the environmentally-related factors affecting flood control should be considered. Not only do infrastructure or other developments on a floodplain expose themselves to risk (depending on their vulnerabilities), but they increase the risk of loss or damage to neighbors or others in downstream communities. Buildings, for example, can increase flood heights and velocities by obstructing the flood flow, reducing floodplain storage capacity, and increasing run-off.

12. Natural events such as fires, windblows or changes in a river's course, affect flood flows. Human activities in the watershed, such as cutting of trees in forestry activities or in clearing for agriculture generally will increase run-off, as will hillside agriculture without adequate terracing or planting on the contour. Paving land in the watershed and on the floodplain will also increase run-off, and installing storm drainage systems will increase the quantity and rate at which rainwaters enter the river system.

Special Issues

Non-Structural Measures or Regulation of Floodplain Use

13. Non-structural flood control measures aim to prohibit or regulate development on the floodplain or watershed areas or floodproof existing structures, to reduce the potential for loss from flooding. As true of any preventative measures, these are less costly than treatment (i.e., installing structural flood control measures). Non-structural flood control measures are essentially environmentally beneficial in that they do not attempt to regulate the natural flooding pattern of a river. The current philosophy among many planners and policy makers is that it is better to maintain undeveloped floodplains as natural overflow areas. Where existing development on a floodplain exists, however, non-structural flood control must be used in conjunction with structural measures.

14. Zoning is an effective means of controlling floodplain development. Zoning the land for such things as agriculture, parks and conservation areas is compatible with floodplain protection, and prevention of land uses which are vulnerable to flood damage. As wetlands have a natural function in flood control, zoning to prohibit activities in wetlands that will reduce their storage flood capability is particularly important.

15. Regulations in zoning ordinances can prohibit or specify the types and function of structures that can be built on the floodway and floodplain to minimize flood risk. For example, the disposal of sewage, toxic and other harmful materials can be prohibited, floodproofing of structures required, and construction of buildings and private roads, that may exacerbate the effects of floods, not allowed.

16. Sanitary ordinances and building ordinances can make further specifications on floodplain management. Sanitary ordinances reduce the risk of health problems which could arise from contamination of water supplies when sewage disposal systems are disrupted by flooding. The ordinances can prohibit the installation of soil absorption systems (e.g., septic tanks, absorption fields, etc.) or require that a permit be obtained prior to installation. Building ordinances can specify structural requirements of new buildings to reduce their vulnerability to flooding, reduce health and safety hazards to occupants (e.g., regulations on electrical wiring and floor elevations), and minimize the extent that the building will impede flow of floodwaters.

17. The ability to apply non-structural flood control measures implies control of land use, and thus is an institutional issue. Non-structural measures can be only as effective as the government is capable of designing and enforcing sound land use.

18. Finally, various actions can be taken which will help decrease or delay runoff and increased infiltration, and thus reduce the risk of flooding. These include watershed management activities (e.g., increasing vegetative cover, particularly on slopes, improving agricultural practices, implementing gully

erosion control measures, etc.), planting of vegetation along river banks to help contain and reduce flooding, and protection of, or restriction of, use of wetlands which have a natural flood control effect.

Social Issues

19. The primary social issue related to flood protection is the unequal distribution of benefits received and costs incurred for the populations affected by flood control measures. In cases where traditional uses of the floodplain for fisheries, agriculture or livestock herding dependent on the natural flooding cycles are disrupted by flood control measures designed to protect other communities (often urban), the rural dwellers frequently are not adequately compensated for the losses incurred.

20. Floodplain dwellers also are those most affected by increased flooding due to changes in land use by others upstream, yet they generally have the least power to enact change or to compel government to intervene on their behalf.

Determination of Floodplain Extent and Frequency of Flooding

21. In order to evaluate flood risk, the probability of floods of various magnitudes on a site must be calculated. The following information is needed:

- the annual peak discharge (maximum rate of flow in a given year) of the river at the site;
- the recurrence interval of various peak discharges (the average interval of time at which the peak is likely to occur); and
- the peak stage or water-surface elevation (or physical limits of inundation) for each particular discharge.

22. As statistical data on peak discharges (a record of a minimum of 10 years is needed) and the extent of inundation with various discharges are often unavailable, historical information from local residents, local records, and geologic analysis of alluvial deposits can be gathered to help determine the periodicity and extent of flooding in an area. Maps then can be drawn up showing areas that are susceptible to flooding. These maps are useful for developing land use plans and regulations for the area.

Project Alternatives

23. Two options may exist for minimizing structural measures which are disruptive environmentally:

- revise operations of existing dams and reservoirs upstream to provide at least partial alleviation of flood risk; and
- use non-structural means, to the extent possible, to reduce flooding risk.

24. When flooding intensity and frequency increases due to man-induced changes in the watershed, non-structural solutions (e.g., revegetating cleared areas, contouring and terracing, tree planting along river levees, reducing access to the area to reduce population influx) can be emphasized. Where flood control is needed to protect existing structures, there may be no alternative but to installing structural protection measures. In this case, the options lie only in the choice of measures, their installation and management to minimize negative environmental impact.

Management and Training

25. The public authority over watersheds and floodplains is divided hierarchically between local, regional and central government, may be shared among government units of equal rank but responsible for different jurisdictions through which the river flows, and finally may be diffused among various functional or operational agencies. The more diffused the administrative structure, the more difficult it will be to achieve the necessary coordination between the technical disciplines and between functions and policies.

26. The adoption of a comprehensive approach to water management calls for a wide variety of professional skills in the planning and policy-making process. In addition to inputs from engineers, economists, and land use planners, contributions are needed from various natural scientists (e.g., geographers, agriculturalists, foresters, livestock and range experts, fisheries experts, ecologists) and social scientists. The involvement of such a large number of specialists poses a challenge for the organization of flood control management.

27. The government unit(s) responsible for flood control must have the planning and regulatory capability to:

- determine the causes, frequency and extent of flooding
- determine the actual or potential effect of various types of development on flood levels
- plan, install or implement required structural and non-structural means for flood control
- implement a flood warning system
- determine areas that are flood prone and take measures to prevent developments that will create a flooding risk
- monitor changes that alter flooding risk and also the effects of flood control measures
- coordinate plans and activities with other agencies responsible for activities in the watershed

28. Where these capabilities are weak, training in technical, administrative, regulatory and community development skills is needed.

29. As already discussed, the ability to apply non-structural measures is largely an institutional issue. The government agencies responsible for flood control must design control measures, particularly non-structural ones, that are appropriate to local conditions. The successful adoption of flood abatement measures often hinges on the interest and support of local communities, and their ability and willingness

to change patterns of land and resource use. A public information campaign may also be a necessary part of a flood control program. In remote areas where the government's influence is weak, local level action can be stimulated by working with traditional power structures and community organizations to encourage necessary change. The government must foster increased coordination and cooperation with these groups when the links are weak.

Monitoring

30. Factors which influence the quantity of water entering and being withdrawn from the river, the land's capacity to absorb floodwater, and the potential damage from floods must be monitored in order to carry out proper planning for floodplain management. Direct and indirect impacts of flood control works should also be closely followed. Data to be collected in routine monitoring should include:

- quantity, intensity, timing and geographical distribution of rainfall and snowmelt
- storm patterns
- soil moisture conditions at various times of the year
- stream discharge (including records of annual peak discharge)
- storage, diversion and regulation of stream flows
- changes in drainage and other factors that affect stormwater runoff
- sediment content of the river water
- sedimentation problems in downstream areas
- changes in the river course and riverbed
- demographic changes in the floodplain and watershed areas
- rural and urban land uses (controlled and uncontrolled land use change on the floodplain and watersheds of the river)
- socioeconomic impacts resulting from the project (including changes to pre-project agricultural, pastoral, fishing practices)
- effects of flood control measures on riverine, estuaries or near-shore marine fisheries
- effects of flood control measures on floodplain vegetation
- effects of flood control measures on wildlands, wildlife habitats and wildlife populations

Table 8.4. Flood Protection

Potential Negative Impacts	Mitigating Measures
Direct	
1. Flooding of lesser magnitude, but greater duration of floodplain downstream due to dam releases.	1. Adaptation by changes in agricultural practices.
2. Potential for structural failure and floodwaters higher than capacity of control structures/measures, leading to increased risk to life and property because local pre-project adaptations are relaxed or abandoned or increased development on the floodplain has occurred post-project.	2. Implementation of non-structural measures to prevent increased flood risk, and of a flood warning system.
3. Cycle of enrichment and groundwater recharge in floodplain soils broken.	3. Where dams are present, partial mitigation of effect by regulation of discharge to imitate natural flooding in a controlled way.
4. Resettlement of populations and other negative socioeconomic effects on populations and communities affected by the project.	4. • Identification of at-risk population groups or groups who may be adversely affected by flood control measures. • Incorporation of their interests and protection into project planning and cost analysis to minimize losses or provide in-kind compensation for losses.
5. Adverse effects on fisheries and other aquatic resources by disruption of migratory routes, deterioration of habitat and changes in water quality (e.g., sediment load), leading to reduced productivity of riverine, coastal and marine fisheries.	5. • Installation of fish passageways. Protection of reproductive sites for fish. • Incorporation of fishery management, including hatchery and re-stocking programs.

Table 8.4. Flood Protection (continued)

Potential Negative Impacts	Mitigating Measures
Direct (continued)	
6. • Negative impacts of channelization measures: • disruption of fish habitat by elimination of pools, riffles and channel irregularities • increased water temperature by removal of vegetation on banks and in stream • increased erosion and sedimentation problems • bed and bank erosion • downstream flooding and sedimentation	6. • Careful selection of engineering options at planning stage. • Limitation of degree of channel modification or maintenance. • Mitigating measures after construction phase. • Minimize reduction of channel length and preserve some meanders. • Limit excavation and fill. • Limit destruction of bank and streamside vegetation. • Replant/reseed banks. • Excavate only one and not both banks, etc. (See Brookes 1988.)
7. Adverse effects of construction.	7. • Minimization of effects by avoiding impediments to natural drainage, uncontrolled run-off and soil erosion, and air pollution. • Provision for adequate filling of borrow areas, control of land clearing, and disposal of spoil. • Limitation of access of vehicles to stream bank.

Table 8.4. Flood Protection (continued)

Potential Negative Impacts	Mitigating Measures
Direct (continued)	
8. Reduction of floodplain grazing, both through ecological changes on the floodplain and intensified development (e.g., irrigated agriculture).	8. • Production of fodder crops and usage of byproducts of irrigated food crops and development of alternative water sources. • Integration of existing rangeland use (e.g., semi-nomadic herding) with planned developments, to ensure substantial grazing and watering possibilities in valley during dry season.
9. Reduction of recession agriculture.	9. Maintenance of natural flooding regime to extent possible in most productive lands (and intensification of production) by maintaining water courses free of flood control structures or installing structures to enable semi-controlled flooding.
10. Obstacles (levees, dikes, etc.) to wildlife passage.	10. Construction of bridges or special crossing places.
11. Loss of wildlands and wildlife habitat.	11. Identification of critical habitats and planning of flood control measures to minimize effects; where habitats or species are dependent on natural flooding regime, minimize disruption of flow in that area to extent possible.
12. Flooding problems created downstream.	12. • Protection of natural overflow areas downstream. • Creation of overflow basins.

Table 8.4. Flood Protection (continued)

Potential Negative Impacts	Mitigating Measures
Indirect	
13. Improved accessibility, development opportunities in floodplain, and sense of security after flood control measures taken, leading to influx of people with associated agricultural development, deforestation, wildlife poaching, infrastructure development, etc.	13. • Limitation of access, if possible. • Planning for anticipated influx and implementation of companion rural development activities. • Introduction of non-structural control measures.
14. Increased fertilizer use on agricultural fields to compensate for loss of fertility, leading to water pollution and dependence on imported supplies.	14. • Optimal timing and rate of application. • Use of nitrogen fixing cover crops. • Use of organic instead of chemical fertilizers.

NATURAL FOREST MANAGEMENT

1. Natural forest management can have several different goals: production of timber and other forest products, watershed protection, and conservation of biological diversity. The focus here will be on projects or project components involved with timber exploitation and the environmental impacts of logging, but management for other products and various agro/sylvo/pastoral systems will also be discussed briefly. The impacts of processing, conservation management, and plantations and reforestation are examined in the following sections of the Sourcebook: "Pulp, Paper, and Timber Processing"; "Wildlands"; and "Plantation Development/Reforestation."

Potential Environmental Impacts

2. Proper management of natural forests can and should support sustainable production of a range of wood and non-wood products, preserve the forests' capacity to render environmental services, conserve biological diversity, and provide livelihoods for various people (including indigenous forest dwellers or tribal peoples who represent endangered cultural assets). Many forest types can provide a continuous supply of timber and other commercial products indefinitely if managed properly. Maintaining an area under forest cover controls erosion, stabilizes slopes, moderates streamflows, protects aquatic environments, maintains soil fertility, preserves wildlife habitat, and provides non-wood forest products important to local economies and households. Sustained harvest of forest products can provide the economic incentive to help prevent conversion of that forest to more destructive land uses and relieve pressure on other forests which are best left undisturbed or under low-impact resource use.

3. If badly managed, however, or cleared for conversion to other land uses, such as agriculture and cattle ranching, a forest can be degraded to secondary forest growth, scrub or wasteland. Poor forest management can increase erosion and siltation of water bodies, disrupt hydrology resulting in increased flooding, water shortages and degradation of aquatic ecosystems, diminish genetic resources, and intensify socioeconomic problems. (See Table 8.5 at the end of this section for a list of all potential impacts and recommended mitigating measures.) The most dramatic impacts arise from clearcutting forests, while the impacts of other activities disrupting ecological processes or changing the character of the forest are subtle yet significant. Large-scale clearing can be a direct or indirect result of forestry (i.e., timber extraction, construction of logging roads) and non-forestry activities, such as infrastructure development (e.g., dams and roads), resettlement programs, and agricultural projects (crops and livestock). Less dramatic impacts can result from low intensity forest land and resource use, including: selective harvesting of various tree species, agroforestry, small-scale livestock operations, fuelwood cutting and collection of other forest products. These activities may not radically alter the quantity of vegetation, or vegetation cover, but can change the quality of the forest by affecting species distribution and composition and ecological processes.

Timber Harvesting

4. The Bank leaves the decision of timber harvesting to others. Specifically, the Bank does not finance logging in tropical forests. The new policy is expected in May 1991. The major direct impacts of timber harvesting result from both the effects of reducing vegetative cover and the physical impacts

of logging operations. The degree of impact will be determined by site conditions (e.g., soil, topography, rainfall), ecological characteristics (e.g., forest type, stand density, wildlife species and numbers present), and harvesting and extraction methods. The following discussion of impacts is necessarily only general in scope.

5. Timber harvesting directly impacts water resources. Surface runoff increases after harvesting, leading to larger and more rapid storm surges in rivers. Decreased infiltration and groundwater recharge, and increased evaporation and storm runoff in the wet seasons affects baseflow and thus lowers streamflow in drier periods. Increased erosion increases sedimentation in rivers and lakes. Stream crossings for logging operations cause direct sediment increases. Felling trees that shade riparian areas increases water temperature. Floating logs down the river for transport, and improper disposal of slash increases the quantity of organic matter entering a stream will alter water quality and can cause oxygen depletion and problems with eutrophication. Fuels, lubricants, pesticides and other substances used in forestry operations can pollute surface and ground waters.

6. Logging also affects climate and air quality. The main localized air quality problems associated with logging operations are dust and smoke. In semi-arid or seasonally dry areas transportation equipment can generate unhealthy quantities of dust, and the soils, exposed after removal of trees and burning, are more susceptible to wind erosion. Smoke from slash burning can cause severe air pollution problems. The accumulation of slash after logging also increases fire. Removal of vegetation results in localized micro-climate changes, and large scale logging operations modify regional temperature, humidity and air circulation patterns. Deforestation, by increasing levels of atmospheric carbon dioxide, a greenhouse gas, significantly contributes to global warming (see "Atmospheric Pollution" section).

7. Land form, slope orientation and grade, and logging practices determine the extent of environmental damage caused by harvesting. Erosion, soil degradation, slope stability and increases in soil temperature are primary concerns in logging operations. The potential for declines in soil fertility after logging is greatest in moist tropical forests whose soils are characteristically nutrient-poor and highly leached. Nutrients are maintained in the intact system by rapid cycling between the vegetation and soil. Dead organic matter is rapidly decomposed and the nutrients taken up quickly by vegetation and soil organisms. Uncontrolled or clearcut logging disrupts this process by removing biomass in which most of its nutrients are stored, and by disturbing soil micro-organisms. Clearing of vegetation cover, thus exposing soils to direct sunlight and higher temperatures, changes populations of micro-organisms and alters decomposition and nutrient transfer. Poorly located or constructed roads on slopes cause land slips, landslides, erosion and sedimentation.

8. The long-term sustainability of tropical timber harvesting is controversial. Deterioration of harvesting sites caused by nutrient depletion and soil degradation (from removal of vegetation and impacts on soil structure and fertility) may not be detectable for hundreds of years in long rotation harvesting systems, so assessing the risk in most existing (selective) harvesting systems is very uncertain. The International Tropical Timber Organization (ITTO) concludes that sustainability is achieved on less than one percent of managed tropical forests. The 1990 Tropical Forest Action Plan (TFAP) review has recommended that forestry projects be halted until sustainability becomes achievable.

9. Harvesting imposes wider ranging impacts on the vegetation than simply the removal of target trees. In the process of logging, other trees and vegetation on the site are damaged or destroyed by falling trees, skidders and other vehicles. The number of non-target trees lost can be higher than that of

harvested trees, particularly in selective harvesting systems. Selective removal of the best trees can lead to genetic erosion of the species in the area. If specimens are not left on the site as seed trees, or the seed trees do not survive the shock of forest disturbance, regeneration of the species is jeopardized. If a cut is large, regrowth will not restore the character of the original forest, at least for a long time. This is particularly true in moist tropical forests where natural regeneration of certain species is problematic. In diverse forests where inter-specific relationships are complex, the removal of some species, even if carried out with low-impact selective harvesting, can have negative impacts on several other species ecologically linked to them. If harvesting creates large gaps in the canopy, windblows can destroy large areas of natural vegetation.

10. Logging in mangrove swamps can be particularly destructive both to the forest itself, which is a finely balanced system sensitive to change, and on adjacent areas protected by the swamp. Mangroves are highly productive coastal ecosystems which offer physical protection to the land from the sea, and protection of in-shore waters from negative impacts from the land (e.g., increased fresh water flow, increased siltation). Timber harvesting in mangrove swamps can be sustainable; but if it is not managed properly, it can result in the destruction of the swamp itself, which is valuable for its wood, fish, crab and shellfish production, and its protective functions.

11. Logging affects wildlife by the destruction of habitat, cutting up of migration corridors, increased poaching pressures, problems of noise and pollution, and hydrologic changes affecting aquatic systems. Again, the impacts can be magnified in moist tropical areas where habitat disruption can set off a ripple effect ultimately affecting a large number of species.

Logging Camps and Logging Roads

12. Construction of logging camps creates a number of environmental problems common to any construction activity, and social problems common to any project involving an influx of population, often of different ethnic, social or economic group from one another and from the local population. The fact that logging of natural forests takes place in remote rural areas where the native population have been isolated from outside forces, heightens the impact.

13. Logging roads create some direct impacts (see "Rural Roads" section), but more important are the indirect effects of such roads. When roads penetrate into remote areas, uncontrolled in-migration of people in search of land for farming or other resources is almost inevitably stimulated. The resulting land use changes often are unsustainable due to the intensity of land use or land uses inappropriate to the environment. Increases in population overload existing infrastructure and social services, such as housing, schools and health facilities, and may cause conflicts over land and resource use rights, racial tensions, and a range of other social problems (see "New Land Settlement" section in Chapter 3).

Management of Secondary Forests

14. Secondary forests, those resulting from cut-over primary forest, could be managed for production and thus reduce pressure on natural forests. These forests generally have better access from population centers than remote areas of natural forest, and may be as productive as plantations without the initial investments. Bringing these forest areas into production can be easier and less environmentally destructive than logging natural forest stands and may be as economically rewarding. Managing secondary forests should be explored as an alternative to harvesting areas that have not been disturbed.

Management for Non-Wood Forest Products

15. Non-wood products represent a largely ignored commercial resource that can generate revenues exceeding those from timber products with a lower capital investment. Latex, oil-seeds, resins, fruit and rattan are non-wood products with well developed markets of high value. Nuts, tannin, medicines, fibers and other "minor forest products", often important in local economies and for household use, may be developed for larger commercial markets. Developing production systems, markets and marketing mechanisms is often difficult; but when successful, it offers a means of sustainable forest utilization with reasonable economic returns and low environmental impact. Loss of the potential of obtaining non-wood products should be considered an opportunity cost of logging natural forests. A potential problem of managing for non-wood products is that once markets are established, demand may rise faster than supply, resulting in the destruction of the resource.

Sustainable Agricultural/Livestock Management

16. Shifting cultivation (also called slash-and-burn or forest fallow) is an ancient form of land use practiced in both moist and semi-arid forest areas. It is a sustainable practice with minimal negative impact on the forest ecosystem if the clearings are small and widely dispersed, and the fallow period between cultivation long enough so that the land can recover. If the fallow periods are too short, a consequence of increased population pressure, the sites deteriorate. Other systems, such as taungya, in which agricultural crops are grown in association with trees until the trees dominate the site, are also successful. Such systems require a stable population so that cultivation does not become intensified to an unsustainable, destructive level.

17. Attempts to integrate forestry with livestock have very mixed results. The quality of the site and intensity of development are critical. Livestock pressure must be low enough that the forest resource is not unacceptably degraded. Clearing tropical forest for cattle ranching in the moist tropical lowland forests, such as in Central and South America, has had disastrous effects, and overgrazing in semi-arid forest areas is a common problem. (See also "Livestock and Rangeland Management.")

External Impacts

18. A number of external forces can result in the deterioration or destruction of the forest ecosystem. These include the flooding of land behind a dam to form a reservoir (see the section on "Dams and Reservoirs"), clearing of forest lands for cattle ranching (see "Livestock and Rangeland Management" section), intensive slash-and-burn agriculture, and conversion to commercial agriculture (e.g., rubber, oil palm, coffee, rice and cacao).

Special Issues

Moist Tropical Lowland Forests

19. The rapid deterioration, if not outright destruction, of many areas of moist tropical lowland forest, which are characterized by high species diversity and ecological complexity, and the difficulties in managing them sustainably are issues of concern worldwide. While conservation of these unique forest areas by establishing parks and reserves is potentially the best way to protect their biological diversity, ecological processes, and the lifestyles of indigenous forest dwellers, only limited areas can be protected

in this way. Economic pressures and population growth are intensifying once-sustainable land-uses (e.g., shifting agriculture) to unsustainable and destructive levels, motivating clearcutting logging operations, and driving the conversion of forest lands for large-scale agriculture and cattle ranching, which generally prove to be unsustainable and permanently damage the forest ecosystem. Management of natural forests for sustainable production of timber and other products to yield important financial returns is one of the most suitable ways to protect forests from conversion to other production-oriented activities while conserving much of their environmental values.

20. The two critical issues in moist tropical forest management for timber production are: (a) the development of sustainable timber management systems, and (b) the implementation of these systems in such a way that the other values of the forest are not reduced to unacceptable levels. In theory, moist tropical forests can provide a continuous supply of forest products indefinitely. In reality, however, few systems exist which have proven to be sustainable or which apply to most of these forest areas. Most of the existing sustainable forest management systems apply to natural forests of a limited number of species. Because of this and economic pressures forcing the generation of quick income from the forest, at present only a very small proportion of tropical moist lowland forests under commercial timber exploitation are managed in a sustainable manner.

21. The most suitable forest management system in moist tropical lowland forests of high species diversity is selective harvesting in which only a small number of trees are extracted per hectare. If done carefully, with minimal damage to the soil and surrounding vegetation, environmental damage can be limited. Impacts on the forest's biological diversity and capacity to provide environmental services are minimized because large spaces in the forest are not created, as is the case with clearcutting.

Social Issues

22. Almost all initiatives that have an impact on natural forests -- whether it be commercial logging, forest-based processing industries, conversion to other uses, other developments (e.g., mining, dam construction, irrigation, industrial development), or closure of forest areas for rehabilitation or conservation -- raise important social issues. Development projects which clear forest areas for other uses can displace people or reduce their access to forest resources on which their livelihoods depend. Commercial logging can destroy resources locally important to subsistence economies and can open up areas to uncontrolled settlement, causing further environmental degradation and social conflict. Similarly, closure of forest areas for rehabilitation or conservation can reduce the income of surrounding populations by depriving them of important foodstuffs or income-generating products. Such closure can lead to greater degradation of adjacent forest areas by people seeking to find substitutes for the resources to which they have lost access. If pressures on the closed area are too great, the conservation or rehabilitation efforts may be unsuccessful.

23. Forest dwellers have considerable knowledge about the qualities, potential utilization, and sustainability of the local flora, fauna, and geologic resources based often on centuries of use. In uplands and arid and semi-arid areas where sources of fodder are often limited, management of forests and local systems of livestock production are usually closely linked; farmers often adopt mixed subsistence strategies in which livestock production based on forest use plays a major role. In the Himalayan region, the productivity of upland agriculture depends heavily on compost and mulching material collected from the forests. Hunting and gathering as well as subsistence slash-and-burn agriculture have been practiced sustainably in moist tropical forest areas for hundreds of years. Artisanal floodplain fisheries are

important to populations in many lowland forest areas. The social organization of traditional groups generally is highly adapted to the demands of particular production systems. Both the technical and managerial knowledge of these resources can be very useful to technical specialists seeking to intensify or modify production in that or a similar area, i.e., adapting agricultural recommendations to areas presently under shifting cultivation, or developing forest management and utilization models for forest lands to be rehabilitated. When forest-dwelling groups are displaced, their indigenous technical knowledge of forest management and utilization is often lost. Careful assessment, including a realistic economic analysis, should be made before assuming that present uses of a forest overgrazing land should be abandoned for something "better".

24. Land tenure issues are almost always of concern in forestry projects. There are often overlapping rights for forest lands, which include state-recognized as well as customary tenure and/or systems of concessionary use rights to products. In the case of forest-dwelling ethnic minorities, there may be strong customary rights to forest lands that may be constitutionally valid despite subsequent transfer of authority over these lands to government. In many societies, rights to land and trees may be held separately, with specific rules for different species. Forest-dwelling groups often have complex tenure rules for forests and products. For example, rights to fruit-bearing trees may be separate from those permitting individuals to put forest land to other uses, including shifting cultivation. Traditional tenure systems may be more appropriate to the management of fragile lands than state-promoted options.

25. Closure of forest areas or restricting access and use of resources affects various population groups differently. For example, landless livestock owners may be most harmed economically by closure of areas, since they, unlike landed farmers, cannot obtain substitute fodder from their own land. Women may have a much greater work burden because of the need to travel further to find substitute resources; yet this burden may not be identified by local people as a problem because of women's lower status in society. Migratory graziers may have their herding routes affected or be forced to overgraze other lands outside the project area which are still available to them, with negative impacts both for those lands and the sedentary groups dependent on them.

26. Planners increasingly are exploring ways to integrate the needs of local people into conservation and forest rehabilitation initiatives through promotion of common property resource management or systems of joint government/user management. It is important to document the existing local management systems, including those which have broken down under increased pressure. In areas of unique biological diversity, other measures have included the creation of buffer zone programs which generate alternatives for the people traditionally dependent on the area to be preserved or the design of conservation systems that allow for controlled utilization of the protected area by local people. A project in Mexico has upheld traditional rights to forests and provided technical support to cooperatives for wood processing industries. The forestry plan for Papua New Guinea has a similar proposal for tribes with traditional forest ownership rights. A Bank project in Nepal involves the extensive rehabilitation of hill forests by strengthening the rights of local user groups to undertake protection and controlled utilization in consultation with local foresters.

Improved Wood Processing Technologies

27. Expansion of the utilization of forest products can help intensify forest management. Many species are not used for lack of processing and marketing infrastructure. In tropical forests of high species diversity, individuals of marketable species are often scattered widely over a large area, making

harvesting difficult and often uneconomical. Logging may also be uneconomical in forests less diverse but in remote areas or of low stocking density. If new products from other species, a wide range of size classes were possible from improvement of processing or development of new markets, a greater proportion of the forest growth could be used. Not only is there a huge scope for developing new products, but conservation of existing supplies (e.g., developing more efficient veneers, wafer board and particle board and through utilization of logging wastes and waste recycling in processing plants) may help balance supply with demand and take pressure off natural forests. The benefits of these approaches are obvious, but so are the dangers. Increased use of a wider range of species may lead to more intensive exploitation of the forests, and if sustainable management systems are not first well developed, could result in large scale clearing or "mining" of the forest resource.

Project Alternatives

28. Alternatives to management of primary (and secondary) forest management for timber, non-wood forest products and extensive, low-impact agriculture and livestock production include:

- reduction of demand for wood through conservation, improved wood stoves and alternative fuels;
- more efficient veneers, wafer board, particle board, use of waste wood and waste recycling;
- increased use of forest species through expansion of processing technologies, and product and market development;
- plantation development to increase production of wood forest products;
- community forestry programs and tree planting by individual land owners for the production of wood products;
- development of eco-tourism as a sustainable economic use of tropical forests;
- encouragement of in-country processing to gain value-added benefits rather than promoting policies which maximize short-term timber harvesting;
- full use of trees destroyed (and often wasted) by forest clearing for non-forestry activities (e.g., dams and reservoirs, road building, industrial and urban development, etc.); and
- intensification of agricultural and tree plantation production on fertile soils or areas already cleared before opening up new forest areas to development.

Management and Training

29. Countries with large areas of natural forest should evaluate their forest resources and develop policies and plans that will protect areas important for their biological resources, watershed functions and cultural resources, and allow for sustainable production of timber and other forest products and for low-impact, sustainable agricultural production (agroforestry and livestock) in the intact forest. In addition, forestry institutions must plan and manage plantations, fuelwood production systems, agricultural activities on degraded forest lands, and community forestry efforts ("Plantation Development/Reforestation" projects are discussed in the next section).

30. The fate of forests depends only in small part on forestry policy. Forestry policy must be integrated and compatible with policies of other sectors that can affect forest lands -- settlement, agri-

culture, energy, industry, trade, infrastructure development and conservation -- and national economic, financial and social policies. Forests are also affected by international factors, chief among them are trade, aid and debt.

31. Driven by the need for income and foreign exchange, government forestry policies in some countries are founded on the principle of maximizing short-term monetary output from the sector. Shortsighted economic assumptions result in over-exploitation of resources, logging in unsuitable areas or improper forest management practices. Many tropical countries with large forest resources have granted harvesting rights to concessionaires for rent, royalty or tax payments representing only a small fraction of the net commercial value of the wood. The problems are compounded by the awarding of short-term leases which impel concessionaires to begin harvesting at once and to clearcut, and by royalty systems inducing loggers to harvest only the trees of highest quality (while in the process damaging or destroying a great many others). There are often no regulations or enforcement of regulations concerning reforestation or minimizing negative environmental impacts from logging. The economic, fiscal, environmental and social costs of these practices can be enormous, with large losses of potential government revenues from forestry and sacrifice of rich biological resources.

32. Trade policies also fuel tropical deforestation. Industrial countries often are permitted to import logs from tropical countries duty-free or with low tariff rates. This acts to discourage them from developing their own timber supplies. In addition, raw logs are often exported from tropical countries. This represents a lost economic opportunity by forfeiting value-added benefits by processing the logs to lumber, poles, railroad ties, plywood, veneer or other products.

33. Forestry projects are dependent on long-term stable institutions that plan, manage and supervise activities. In many developing countries, however, forestry departments need strengthening; they are of low priority and suffer from a lack of trained personnel, shortage of operating funds, and weak research, extension and monitoring capabilities. Forestry training institutions are often absent or neglected. Forestry data generally is limited. Foresters, traditionally trained to protect forest reserves and manage them for industrial production, are often not prepared to manage forest stands for non-industrial use or deal with the socioeconomic aspects of forest management.

34. Government forestry institutions must have the capability to:

- establish forestry policy
- coordinate with central government agencies responsible for other sectors and national policies which will affect forest resources
- prepare forest management plans
- undertake forest inventories and silvicultural research
- regulate and tax exports and establish markets for forest products
- designate, finance and oversee the management of parks and preserves
- develop training, extension and public awareness-raising programs
- plan transportation and other infrastructure related to forestry operations
- ensure that population groups and local communities are adequately represented, and incorporate them into decision-making processes
- promote appropriate technologies which stimulate local economies in the region and employ local labor
- coordinate and cooperate with non-governmental organizations

35. To increase the technical depth of and planning and management expertise of forestry staff, training may be needed in the following areas: (a) policy, regulation, marketing, economics, management, organization, accounting, personnel, contracts, evaluation, reporting, and conflict resolution; (b) technical skills in mapping, biological surveys and inventories, forest sciences, forest management, forest engineering, and environmental impact assessment; and (c) research, education and extension.

Monitoring

36. Monitoring in forestry projects is extremely important to determine that management plans are being followed and that the forest stand treatments are achieving the desired results. Monitoring should ensure that:

- loggers, harvesters and road builders adhere to conditions set forth in their contracts designed to minimize environmental impacts;
- harvesting and transportation do not create unanticipated environmental problems (monitoring of soil erosion, soil fertility, stream water quality, groundwater level, vegetation and wildlife changes);
- changes in species and site conditions are identified and stand treatment prescriptions are altered as appropriate;
- only designated areas are accessed and only the specified species and volumes are harvested;
- natural regeneration after harvesting occurs as predicted (rate of cover restoration, rate of regeneration of various species);
- objectives of the overall development project are being met and infrastructure to regulate and manage the project is adequate;
- no unpredicted socioeconomic impacts occur and if so adequate steps are taken to mitigate them, and that a mechanism exists for community organizations to monitor and evaluate the project and voice their concerns on a regular basis; and
- financial distributions are legal and according to contracts and these are adjusted in the event that social services are over-burdened or costs exceed predicted values.

37. Frequency of monitoring is dependent on site conditions, size and complexity of the project. Actual harvesting should be monitored at least weekly by a trained forester. A standard procedure for log inventory should be continuous, and regular auditing of revenues should be done by an impartial company. A periodic reassessment of environmental impacts should be done, with recalculation of costs and benefits and appropriate changes in the management plan should be made.

38. In addition, changes external to forestry operations that affect the forestry resources must be monitored. This includes environmental affects of other development activities and natural events as well as demographic changes in forest areas.

Table 8.5. Natural Forest Management

Potential Negative Impacts	Mitigating Measures
Direct: Logging	
1. • Soils • Erosion: disturbance of the forest understory and soil, increasing susceptibility to water erosion. • Slope stability: roadcuts across sloping terrain and clearing of vegetation on the slopes resulting in landslips and landslides. • Nutrient loss: loss of nutrients from timber harvesting and increased leaching; exposed, disturbed soils where vegetation has been removed. • Temperature: dramatic increase in temperature after removal of forest canopy, killing soil organisms or drying the soil to extent that regeneration is hindered. • Structure: compaction and loss of organic matter altering soil structure, and reducing infiltration, water holding capacity, aeration and root penetration; laterization.	1. • Avoid logging in the rainy season and establish criteria for logging on slopes and near water; and clearly mark areas that should not be harvested. • Supervise logging to reduce damage and encourage rapid regeneration. • Use low impact harvesting equipment and methods and minimize skid trail distances. • Locate log landing in well drained, easily accessible areas downslope so a straight skid road can be followed. • Restore land by grading and reseeding disturbed areas, including guidelines, locating them away from slopes and water and keeping them well maintained. • No whole-tree harvest in areas of low nutrient levels, leaving all but boles on the site.
2. • Vegetation • Species composition: species diversity decreased by selectively harvesting the best stems of valuable species. Soil conditions and light regimes created by various harvesting methods influencing regeneration dynamics of forest stands.	2. • Collect information or sponsor research on plant community dynamics, regeneration biology and silviculture of forest type. • Consider (and perhaps research) various regeneration and harvesting methods.

Table 8.5. Natural Forest Management (continued)

Potential Negative Impacts	Mitigating Measures
Direct: Logging	
2. • **Vegetation (continued)**	
• Weeds: opening of the forest canopy resulting invasion of weeds, impeding natural regeneration and reforestation efforts. • Slash: logging debris as a fire hazard and impediment to regeneration. • Blow downs: increased danger due to opening up patches by logging.	• Collect information or sponsor research on plant community dynamics, regeneration biology and silviculture of forest type. • Consider (and perhaps research) various regeneration and harvesting methods. • Choose silvicultural system that will ensure regeneration and sustainable production and minimize damage (leave adequate number and quality of seed trees, selective harvest, small cuts to avoid large gaps). • Establish preserves/parks of ecologically significant forest areas, ensuring that area is large enough to maintain biological diversity, ecological processes and cultural assets.
3. • **Wildlife**	3. • Maintain inventory of and collection of research results on species present in the area.
• Fisheries: sedimentation, nutrient loading, changes in streamflow and water temperature caused by logging causing dramatic changes to fish populations. • Wildlife habitat: disruption of habitat, loss of tree species on which wildlife species depend, and disturbing migratory routes of wildlife leading to depletion of wildlife. • Presence of machines and people: disturbance of wildlife through logging and transport activities. • Poaching: increased poaching of wildlife due to influx of people resulting directly and indirectly from forestry activities.	• Plan harvesting intensity, methods and timing based on this information. • In particular, ascertain presence of or migratory use by endangered species through contact with wildlife professionals in government, NGOs and universities.

Table 8.5. Natural Forest Management (continued)

Potential Negative Impacts	Mitigating Measures
Direct: Logging (continued)	
4. • Air	4. • Reduce burning.
• Dust: logging activities and timber transportation on dirt roads generating large amounts of dust in dry season or semi-arid zones.	• Avoid creating large tracts of open land. • Limit operations when dust and fire are a problem and plan transportation routes to avoid population centers.
5. • Water	5. • Maintain vegetation as buffer zones adjacent to every body of water.
• Extremes of flow: soil infiltration and water holding capacity of harvested areas reduced such that water runoff is much more pronounced, aggravating flooding when rains come and low flow during dry season. • Groundwater recharge: rain water recharge of aquifers reduced due to increased runoff. • Ponding and stagnation: land form changes, water course obstruction and soil compaction causing localized ponding and stagnation. Increased sedimentation altering natural aquatic biology and water course features. • Increased temperature: opening of the forest canopy adjacent to water body increasing water temperature, which alter aquatic biology and water chemistry. • Contamination: water pollution from petroleum products, herbicides and organic waste associated with forestry operations. Turbidity due to increased sediment loading altering sunlight penetration, affecting aquatic plants and damaging fish species.	• Assess impact of forestry on sediment and nutrient loading on stream flow and methods used to reduce impact. • Provide adequate waste disposal facilities. • Procedure for use and storage of chemicals, oil and fuel to minimize potential for pollution.

Table 8.5. Natural Forest Management (continued)

Potential Negative Impacts	Mitigating Measures
Direct: Logging (continued)	
6. • **Social and Cultural** • Local economic and social customs: impacts on labor market and labor availability for food production, a shift to more cash-based economy, alteration of daily living patterns and political power structure changes are common. • Land tenure and traditional forest uses: hunting, gathering and traditional exploitation of forest resources disrupted; limitation of access of resources by local populations. • Overloading of infrastructure and social services (e.g., housing, education and health services) by in-migration of forest workers and spontaneous settlers and social problems such as an increase in crime, alcoholism, disease and violence.	6. • Incorporate local communities in planning and execution of project. • Develop local infrastructure to handle increase in population (i.e., waste disposal, school, health care and law enforcement). • Protect significant cultural landmarks and traditional land and resource use patterns. • Establish clear, long-term jurisdiction over the forest emphasizing local involvement in decision making. • Involve local leaders in protection to avoid illegal harvesting or settlement. • Monitor and control disease.
Indirect: General	
7. • Increased access: roads opening forest areas causing uncontrolled population in-migration with subsequent problems.	7. See "Rural Roads" section.
Indirect: Road Construction and Transport	
8. • Direct impacts (e.g., increased soil erosion and sedimentation of surface water) and indirect impacts of road construction (see "Rural Roads" section).	8. Align route, drainage works, etc. (See No. 7.)

Table 8.5. Natural Forest Management (continued)

Potential Negative Impacts	Mitigating Measures
Indirect: Road Construction and Transport (continued)	
9. • Degradation of existing public roads by heavy timber loads.	9. • Restrict load size. • Use road taxes to upgrade road.
External	
10. Cattle-ranching operations that clear forests for pasture.	10. See "Livestock and Rangeland Management" section.
11. Conversion to commercial agriculture (rubber, oil palm, coffee, rice, etc.).	

PLANTATION DEVELOPMENT/REFORESTATION

1. This category includes projects and project components which plant trees for production or for environmental protection. Forestry products include lumber, pulpwood, poles, fruit, forage, fibers and fuel. The scale of production ranges from large commercial plantations, to community woodlots, to individual farmer plantings around houses or farm fields. Protection-oriented activities include planting trees for slope stabilization, sand dune fixation, shelter belts, various agroforestry systems, living fences and shade trees.

Potential Environmental Impacts

2. Plantations, reforestation on degraded lands, and social forestry plantings have a number of positive impacts resulting from goods produced and environmental services rendered.

3. Plantations offer the best alternative to logging natural forests as a means of satisfying demand for timber and other wood products. Plantations for timber production generally use faster growing species, and are easier to access and harvest than natural forests producing more uniform, marketable products. Similarly, community-oriented plantations for production of fuelwood and forage located near settlements can both facilitate users' access to these goods and at the same time help alleviate pressure on local vegetation leading to overcutting and overgrazing. Generally established on lands unsuitable or marginal for agriculture (e.g., on existing forest lands or on degraded areas), tree plantations represent a beneficial and productive land use which does not compete with uses of higher productivity.

4. Reforestation provides a range of environmental benefits and services. Reestablishing or increasing tree cover can increase soil fertility by improving moisture retention, soil structure, and nutrient content (by decreasing leaching, providing green manure, and adding nitrogen if nitrogen-fixing species are used). Where shortages of fuelwood have caused the use of animal dung for fuel instead of fertilizer for agricultural fields, fuelwood production indirectly will help maintain soil fertility. Tree planting stabilizes soils by reducing water and wind erosion on slopes, in adjacent agricultural fields, and on unconsolidated soils such as sand dunes.

5. Establishing tree cover on bare or degraded lands helps reduce rapid runoff of rainwater, thus regulating stream flow and improving water quality by reducing sediment inputs into surface waters. Cooler temperatures and moderated wet and dry cycles under trees provide a favorable microclimate for soil microorganisms and wildlife, and can help prevent soil laterization. Plantations have a moderating effect on winds and help settle dust and other particulates out of the air. Incorporating trees into agricultural systems can improve crop yields, by virtue of their positive effects on the soil and climate. Finally, the increased vegetation cover established by large scale plantation development and tree planting projects represents a carbon sink, a short-term answer to global warming caused by carbon dioxide build-up in the atmosphere.

6. Large commercial plantations have the potential for causing negative environmental impacts of considerable scope and magnitude. The worst impacts occur where natural forests are cleared for plantation establishment. Apart from systems practicing enrichment plantings or underplanting, the ground identified for a plantation generally is prepared by clearing competing vegetation. The negative impacts

of site preparation includes not only the loss of existing vegetation and the ecologic, economic and social values it may have, but also environmental problems associated with land clearance: increased erosion, disruption of the hydrologic cycle, compaction of the soil, loss of nutrients, and a resulting decline in soil fertility. Although destructive, many of these impacts can be short-lived; the site starts to recover once vegetation cover is reestablished by replanting. (These impacts are summarized in Table 8.6 at the end of the "Plantation Development/Reforestation" section.)

7. Plantations are man-made forests whose trees are treated essentially as long rotation agricultural crops. As such, many of the negative environmental impacts inherent in agriculture are present in plantation forestry. The degree of impact depends in large part on site conditions prior to planting, site preparation techniques, the species planted, treatments during the rotation, length of rotation, and harvesting methods. Reforestation and afforestation activities, particularly in drier regions, can deplete soil moisture, lower water tables, and affect base-flow into streams.

8. Like any agricultural crop, plantations of fast growing, short rotation trees can deplete soil nutrients and decrease site fertility by repeated removal of biomass and disturbance of the soil. Although this also can be true of long rotation species, the effects are less obvious. Soil compaction and damage occur during site clearing (removal of vegetation through physical means or burning), mechanical site preparation and harvesting. Erosion can occur in plantations with incomplete canopy closure or limited undergrowth. Accumulation of leaf litter under plantations increases the risk of fire and decreases rainwater infiltration, and the dominance of one or two species in the leaf litter can change chemical and biological characteristics of the soil. Leaf litter of coniferous (e.g., pine) plantations can acidify the soil.

9. Some species are allelopathic, producing toxins which inhibit seed germination of other species. Irrigated plantations can compete with other demands for water and cause other environmental and social impacts common in irrigation projects (see "Irrigation and Drainage" in the next section). The return water from irrigated plantations in semi-arid areas may be saline, rendering it less useful for other purposes and degrading the quality of surface waters into which it feeds. Chemicals (pesticides and biocides) used to control insect pests and diseases, and fuel and oil used in forestry equipment can pollute surface and ground water and can be a direct health hazard for those that use them.

10. The indirect impacts of large commercial plantations include those resulting from road building for wood transport (see "Rural Roads" section) and from wood processing industries (see "Pulp, Paper, and Timber Processing" section in Chapter 10).

11. Tree planting as part of a social forestry program can take various forms, including village or community woodlots, planting on government owned land or along rights of way, and planting around agricultural fields, along streams and beside houses. There are few negative environmental impacts associated with such plantings. The trees provide useful products, and environmental and aesthetic benefits. The common problems occurring in these activities are social in nature (see discussion below).

12. Tree planting for protection purposes -- e.g., shelter belts, windbreaks, slope stabilization, erosion control, watershed management, stream bank protection, sand dune fixation -- also are beneficial in essence, carried out for the purpose of providing environmental protection and services. If problems arise they are most likely to be social (e.g., land and resource tenure issues).

Special Issues

Use of Exotics

13. Plantations and conservation plantings are often established with exotic rather than native species. This is done because (a) the exotics chosen are faster growing than the native species or have more desirable characteristics for its end-use; (b) seed of the exotics is readily available through commercial sources; or (c) the growth and end-use characteristics of the exotic species are better known than those of the native species. Using exotics in an area for the first time always carries a risk. While the use of exotics has been very successful in many places, it has caused problems or unrealistically high expectations in others. An exotic introduced into a new environment does not always do as well as anticipated. This can be due to unsuitable site conditions at the margins of the species' ecological tolerance (i.e., rainfall, temperature); attacks of pests or diseases (sometimes devastating) to which the species has little or no resistance; and inadequate site preparation, planting and subsequent plantation maintenance.

14. Native species, although often more slow growing than exotics, generally have a longer-term viability; genetically selected and refined over the centuries, they are adapted to the ambient conditions and are thus better able to survive local climatic extremes and outbreaks of pests and disease. There have been cases where "slow-growing" native scrubland vegetation has been cleared at considerable financial and social cost and replaced with plantations of "fast-growing" exotics, which ultimately had lower productivity than the cleared native vegetation or than was justified by the costs. There have also been cases where the unpredictability of the species' performance in a new environment has resulted in the opposite phenomenon: overproduction. Exotics can become weeds, spreading from the planting site over the landscape, taking over in areas they are not wanted and becoming almost impossible to eradicate.

15. An exotic species (and various provenances of that species) should be well tested before it is used extensively in a new area. Particularly in protection plantings where rapid establishment and maintenance of forest cover are critical, fast-growing exotics should be interspersed with native species, which although may be slower growing, are more reliable on the long-term. Much more research is needed on the botanical and ecological characteristics and end uses of the many little-known tropical species. Much information can be gained through local peoples who are knowledgeable about the native vegetation.

16. An additional problem is that an exotic may not be acceptable locally for the end use for which it was planted (e.g., polewood, firewood). A species used widely in one place for fuelwood, for example, may not be suitable in another with different foods and cooking requirements. People's perceptions about the qualities of wood and other forest products can be culturally based and strongly imbedded. Prejudices against a species, for whatever reason, may die hard. Before a species is introduced into an area, its local acceptability must be tested.

Socioeconomic Issues

17. The socioeconomic benefits of large commercial plantations include employment generation (more jobs than in natural forest management but less than in agriculture) and often some improvements in local infrastructure and social services. As with timber harvesting operations in natural forests (see "Natural Forest Management" section), however, these have negative aspects, particularly in remote areas. These are problems related to the labor force brought in from the outside (overloading existing local

infrastructure and social services, social and perhaps racial tensions, increased health problems, etc.); increased monetization of the economy; and, if the plantations result in the construction of roads, problems with unplanned population influxes and societal changes stemming from greater exposure to the outside world.

18. Local reforestation efforts, such as community woodlots and plantings around homesteads, can have many direct benefits for individuals and communities. Fuelwood projects can greatly reduce the time and effort involved in gathering fuel for cooking, making this time available for other activities. Planting for forage production can improve access to and availability of livestock feed which is particularly important in drier periods. Significant income can be derived from sales of wood, fruit, nuts, fibers or other products from the trees. The timing of tree cutting is somewhat flexible so it can be done to coincide with favorable market conditions or a time when the wood or cash is needed most. Community woodlots can provide short term employment for landless and poorer people in the community, mainly at the stages of plantation establishment and harvesting. The relatively low labor and capital requirements of trees between the time they are planted and harvested are advantages to farmers planting trees on their own land. Because trees can be grown on marginal land unsuitable for farming or on small areas of unused land, they do not compete with other more valuable land uses.

19. Sizable plantings, either privately owned plantations for commercial timber production, community plantations for production of fuelwood or other products, or large scale protection plantings (e.g., watershed management and sand dune stabilization) can pose problems related to land tenure, and land and resource use rights. Programs planting on communal land often overlook or are unaware of the traditional rights of land use or right-of-way. Even protection plantings on degraded lands can cause social conflict. Even if land is degraded (and perhaps causing it), local people may be using it for fuelwood or fodder cutting, livestock grazing, or as a route for themselves and livestock. Planting trees on this land and restricting people's access to it, even if it is in theory to benefit these people, will cause local hardship if a suitable alternative is not provided in compensation.

20. A common mistake of plantation and reforestation projects is to overlook the diversity of wild foodstuffs (mushrooms, roots and tubers, greens, fruits and honey, nuts, condiments, edible oils, etc.) found in forests, grasslands, and uncultivated scrub areas along roads or near cultivated fields -- and collected and marketed mostly by women. Particularly in arid and semi-arid areas, these foodstuffs can be key elements in household nutrition or as income-earners in times of drought. Wild foodstuffs tend to be considered food of lower status as economies become more monetized and urbanized, or people more formally educated. The opportunity to increase the production and utilization of such products and the positive impact on food security is generally overlooked, even in plantation forestry projects that profess poverty alleviation and sustainability objectives. An environmental assessment needs to collect data on the availability and use of such foodstuffs over time by specific ethnic and economic groups in the project area and bring this information to the attention of planners.

21. Not only existing uses of resources, but future possibilities to generate a wider range of products, are overlooked. A common mistake in plantation projects that seek to take pressure off forest reserves is to focus plantation production on too-limited a range of products for local needs, with the result that people continue to exploit the forest areas. Production of raw materials for local income-generating enterprises is seldom attempted in large-scale Bank projects, because it is perceived as requiring too much institutional development to coordinate the needs of dispersed enterprises and therefore of low economic returns.

22. Ownership of the trees and the land on which they are grown is often a problem. In many countries all trees, planted or natural, officially belong to the government, which is a disincentive for reforestation. Also, planting on what is thought to be communally owned land, but which is in fact land to which certain people have rights, means that the products designed to be for the benefit of community members will be claimed by a limited number of people, often not the most needy.

23. A number of socioeconomic problems arise in "social forestry" projects because of the relative newness of these types of activities. Often neither government forestry personnel nor local people are well versed in the skills required. Participatory social forestry activities must involve rural peoples in organized activities for producing and managing their trees or forest stands for their own benefit. Their participation cannot be assumed; the benefits and costs of participation must be clear to them and the obtainment of those benefits due them unencumbered. A profound behavioral change is called for in switching from the traditional gathering of fuelwood and other forest goods from natural forests to the cultivation of trees themselves.

24. Forestry personnel, generally trained in natural forest management and often charged with guarding forest stands from the people, often lack the confidence of the rural people as well as skills in communication and social systems analysis needed for the social forestry job. As a result many problems arise. Species are frequently planted because of seed availability and not end use or site suitability. Once plantations are established people do not know how to care for them, when to thin and harvest, or what to do with the trees when they are harvested. Plantations initially intended for one purpose may eventually be used for another purpose, or not used at all because there has been no clear plan for their use. Often people who have spent generations cutting down trees to increase agricultural land do not understand the benefits of forestry projects that take land out of agriculture. Equipment for harvesting and transporting trees is often lacking.

25. Finally, there are a number of economic risks associated with plantation forestry. Markets for forest products are volatile, or may disappear during the long life of a single rotation. Political and economic conditions may change, which will alter priorities and allocation of funds. Fire, insects and disease can destroy an entire tree crop. If the market drops because of low demand, increased transportation costs, or flooded markets, there can be a net loss. Enthusiasm in a forestry project whose tangible benefits will not materialize for a relatively long time (a minimum of 3 years), will be dampened by the short-term needs of people.

Project Alternatives

26. The following alternatives to plantation forestry exist: (a) devising management systems for natural forest or shrubland as an alternative to clearing for plantations whose direct costs (establishment and maintenance) and indirect costs (opportunity costs of precluding other land and resource uses) may not be justified; (b) increasing efforts to conserve wood resources (e.g., by promoting more efficient woodstoves, more efficient veneers, particle board, etc., and use of waste wood and recycling), and to promote alternative fuels (e.g., natural gas) as effective alternatives to fuelwood plantations; and (c) where tree planting is being considered for stabilization of slopes or land rehabilitation and improvement, protection of the land by fencing or guarding (to prevent livestock grazing and encourage colonization by native species) as an effective and inexpensive alternative to tree planting.

Management and Training

27. Whether a private plantation of cash crop trees, a community woodlot project or a large-scale commercial plantation, institutional support and training throughout the process is important. (For a discussion on general issues concerning forestry institutions, see the "Natural Forest Management" section.) Typical management support will include:

- project planning, implementation and evaluation
- work planning, budgeting and financial management, personnel management
- development of community education and extension programs
- procurement of seed and equipment
- handling of land tenure issues and enlistment of support of local communities and political leaders
- development of markets and local processing capabilities

28. Technical support and training may be needed in:

- site assessment and mapping
- selection of species
- management of propagation material (seed: purchase or collection, storage and handling; vegetative material [i.e., cuttings]: collection and handling)
- nursery operations
- planting methods and timing
- management techniques, such as thinning, coppicing, fertilizing, irrigating, monitoring insects and disease, weeding, and protection against livestock damage and poaching
- collection of growth and yield data to evaluate stand performance
- harvesting and transportation
- operation of equipment
- development of end uses for the wood and by-products of the plantation operation
- marketing procedures

29. Forestry personnel must have a range of new skills to be able to support social forestry efforts. These include:

- identification of specific groups which could be involved in tree planting
- communication and extension skills
- technical knowledge about establishment and management of trees suited to various social forestry activities
- land and resource tenure issues

Monitoring

30. The following factors should be monitored:

- environmental impacts of site preparation and replanting quality of planting stock
- growth rates of the plantation
- weed problems
- presence of pests and disease
- management treatments: if being done properly and according to schedule
- protection of the stands
- market trends
- distribution of revenues and benefits from the plantations
- changes in costs and benefits as conditions change
- recalculation of costs and benefits as conditions change
- pressure on agriculture, land tenure, natural forests
- environmental impacts of harvesting
- long-term viability of the plantation from ecological, economic and managerial standpoints

WATERSHED DEVELOPMENT

1. Watershed development projects are based on the premise that hydrologic units are appropriate for conceptualizing strategies and planning investments. They display a complete range of intersectoral linkages. Failure to approach planning on a watershed-wide basis can lead to incorrect priorities, illogical sequencing of interventions, and interference between separate activities in different sectors or even in the same sector. The emphasis in watershed development has generally been on agriculture and forestry projects, but should extend to any intervention involving land uses which affects the hydrologic cycle. Many watershed projects have concentrated on the technical aspects of investments to correct problems caused by deforestation, declining soil fertility and productivity, erosion and sedimentation, and flooding and drought. A recent review of such projects in Asia has shown that political, social, economic and institutional issues must be addressed as well to ensure success.

Potential Environment Impacts

2. Watershed development projects or project components are undertaken for purposes which are consistent with sound environmental management. They are unlikely to fall into Category A or even B and thus do not require a formal EA. Consequently, they are not discussed further in the Sourcebook. The user is invited to read the section on "Land and Water Resource Management" in Chapter 2 for more information on the watershed approach to environmental planning and management.

Table 8.6. Plantation Development/Reforestation

Potential Negative Impacts	Mitigating Measures
Direct: Site Preparation	
1. Soil erosion from clearing site.	1. • Reestablishment of forest cover as soon as possible after clearing. • Use of fast growing, intermediate tree crops or mulching of exposed soils. • No clearing on steep, unstable slopes or highly erosive soils. • Limitation of plantation size or stand sizes. • Limitation of site preparation to dry season.
2. Soil compaction and puddling by machinery.	2. • Limitation of use of machinery. • Manual site preparation.
3. • Loss of organic matter and nutrients by removal of vegetation and leaching. • Development of hardpans and laterization.	3. • Rapid replanting. • Cover crops. • Mulching.
4. Where burning is involved, air pollution from smoke.	4. • Limitation of use of fire and size of burn where possible. • Burning in wet season.

Table 8.6. Plantation Development/Reforestation (continued)

Potential Negative Impacts	Mitigating Measures
Direct: Plantation Management and Harvesting	
5. Soil erosion from harvesting.	5. • Replanting as soon as possible after cut. • Avoidance of clear cutting; practice of "small coupe logging" (characterized by checkerboard pattern of alternating small cuts with unlogged areas). • Limitation of harvesting to dry season or season of low rainfall. • Planning of felling to minimize log skidding and avoidance of skidding logs parallel to slope. • Stabilize skid trails as soon as possible after use. • Use of animals instead of skidders for extraction.
6. Loss of nutrients from the system by thinning and clear cutting and by whole-tree harvest.	6. • Logging debris left on ground after harvesting and removal of boles only (no whole-tree harvesting). • Planting of cover crops between rotations; addition of fertilizer to compensate for nutrients loss.
7. Use of fertilizer, pesticides and herbicides having negative impacts on-site and on quality of local water bodies.	7. • Limitation of potential of pest and disease infestations by choice of resistant species. • Choice of chemicals with least potential negative impacts. • Controlled use of chemicals.

Table 8.6. Plantation Development/Reforestation (continued)

Potential Negative Impacts	Mitigating Measures
Direct: Plantation Management and Harvesting (continued)	
8. Chemical and biological changes in the soil as litter becomes dominated by one or a few species and decomposition dynamics are altered.	8. Limitation of size of stands and interspersal with stands of native vegetation.
9. • Direct damage in harvesting operations by dragging and skidding logs causing compaction.	9. • Use of manual methods or animal power for clearing forest instead of mechanical means.
• Localized soil erosion and unequal distribution of debris and organic matter over the site.	• In short rotation plantations plan use of same tracks and loading areas in harvesting operations to protect as much of site as possible.
10. In semi-arid zones depletion of soil moisture and lowering of water table in plantation area.	10. • Choice of low water demanding species. • Water catchment and conservation techniques to minimize runoff and evaporation losses and maximize infiltration.
11. Build up of organic matter under plantations posing a fire hazard.	11. Periodic clearing or burning to keep volume low.
12. Increased sedimentation of streams.	12. • Buffer zones of undisturbed forest 20-40 m wide along streams. • Avoidance of earthfill dams across streams as crossings. • Sediment traps in streams. • Avoidance of skidding trees in stream.

Table 8.6. Plantation Development/Reforestation (continued)

Potential Negative Impacts	Mitigating Measures
Direct: Plantation Management and Harvesting (continued)	
13. Increased organic matter entering surface waters in form of leaf litter and logging debris or from logs transported on river leading to decrease in water quality and perhaps eutrophication, and navigational hazards.	13. • Buffer zones along streams. • Spaced transport of logs in river over time.
14. Soil erosion from logging roads.	14. • Siting of roads on ridge tops or valley bottoms and avoidance of steep grades on hillsides. • Engineering to ensure proper drainage or provision of drainage measures. • Stabilization of road cuts with mulch, wood chips, etc. • Minimized use of borrow pits or stabilization after use. • Proper road maintenance. • Use of rivers for log transport (see also "Rural Roads" section).
Displacement of Other Ecosystems	
15. Loss of habitat and decreased biological diversity by replacement of natural forest by plantations with limited number of species and increased uniformity of forest structure.	15. • Protection of natural forest area with particularly high or unique biological diversity. • Limitation of plantation establishment to degraded sites or sites of low diversity.

Table 8.6. Plantation Development/Reforestation (continued)

Potential Negative Impacts	Mitigating Measures
Displacement of Other Ecosystems (continued)	
	• Increase in number of species planted and avoidance of monocultures over large area. • Restriction of size of individual stands and mixing of stands of various age classes. • Conservation of islands of untouched forest or natural vegetation. • Separation of stands by belts of native vegetation and use of native species as plantation species.
16. Increased potential for massive loss by pests or pathogens (through simplification of natural ecosystem, provision of abundant food for pest, increased pest habitat, absence of natural controls (e.g., in the case of introduced exotic tree species).	16. • Use of native species. • Choice of species and provenances with pest or disease resistance. • Rotation length to minimize susceptibility (e.g., cutting before trees are overmature). • Thinning and other stand improvement measures to remove dead and diseased material, and wood residues which act as centers for infection. • Direct pest or disease control.
17. Loss of forest products from native species.	17. Careful evaluation of local use of forest products to accommodate continued use and determine feasibility of developing local industries based on these goods.

Table 8.6. Plantation Development/Reforestation (continued)

Potential Negative Impacts	Mitigating Measures
Displacement of Other Ecosystems (continued)	
18. Spread of plantation species outside of plantation becoming a nuisance, competing with native species and becoming weeds in agricultural fields.	18. Species choice to avoid ones that will grow out of control from desired site.
Socioeconomic Impacts	
19. Social impacts from influx of people from outside, both wage earners and spontaneous setters, induced by road building into remote areas (direct and indirect impacts).	19. See the following sections: "New Land Settlement"; "Indigenous Peoples"; and "Induced Development."
20. Problems related to land tenure and land and resource use rights leading to unequitable sharing of costs and benefits of the project.	20. • Genuine integration of local communities and peoples in project planning and implementation. • Pre-project socioeconomic surveys and assessments and land and resource use studies. • Provision of alternatives which fairly compensate local people who incur losses.

IRRIGATION AND DRAINAGE

1. Irrigation and drainage projects manage water supplies for the purpose of agricultural production. There is a wide variety of irrigation types depending upon the source of water (surface or groundwater), means of water storage, conveyance and distribution systems, and methods of delivery (field application).

2. Large scale utilization of surface water (predominantly rivers) for irrigation has long been practiced, in some countries for thousands of years, and still accounts for the major public sector investments in irrigation. Large scale irrigation projects using groundwater are recent phenomena of the last thirty years. They are found mainly on the great alluvial basins of Pakistan, India and China where the tubewells tapping into the groundwater are often used in conjunction with surface water irrigation systems.

3. The dominant delivery method (up to 95 percent worldwide) is surface irrigation (flood or furrow irrigation) in which water is distributed over the irrigated area by gravity in overland flow. The other systems are sprinkler and drip (trickle) irrigation. Sprinkler irrigation involves spraying water droplets over the land surface in simulation of rain. In drip irrigation water is released in drops or a light stream through holes in plastic tubing laid on or buried below the surface of the soil. Although they are relatively new technologies requiring higher initial investment and more intensive management than surface irrigation, sprinkler and drip irrigation show great potential for maximizing the efficiency of water use and reducing irrigation-related environmental problems.

4. Irrigation projects can include the following facilities and infrastructure: (a) dams, watershed and reservoirs; (b) diversion and intake facilities; (c) wells, pumping stations, canals, ditches and pipelines for the conveyance of water (including drainage); and (d) distribution systems for sprinkle and drip irrigation.

Potential Environmental Impacts

5. The potential negative environmental impacts of most large irrigation projects include: waterlogging and salinization of soils, increased incidence of water-borne and water-related diseases, resettlement or changes in the lifestyle of local populations, and increases of agricultural pests and diseases resulting from the elimination of dry season die-back and the creation of a more humid microclimate. The expansion and intensification of agriculture made possible by irrigation has the potential for causing increased erosion; pollution of surface and groundwater from agricultural biocides; deterioration of water quality; and increased nutrient levels in the irrigation and drainage water resulting in algal blooms, proliferation of aquatic weeds and eutrophication in irrigation canals and downstream waterways (see Table 8.7 at the end of this section for a summary of all potential impacts). Increased quantities of agricultural chemicals are usually required in irrigated lands to keep production levels up; fertilizer must be used to compensate for high growth rate and loss of nutrients through leaching, and pesticides to control the greater numbers of crop pests and diseases.

6. Large irrigation projects which impound or divert river waters have the potential to cause major environmental disturbances resulting from changes in the hydrology and limnology of river basins. (The

effects of water impoundment are discussed in the section on "Dams and Reservoirs.") Reducing the river flow changes floodplain land use and ecology, disrupts riverine and estuarine fisheries, and causes salt water intrusion up the river and into the groundwater of adjacent lands. Diversion and loss of water through irrigation reduces the water supply for downstream users, including municipalities, industries and agriculturalists. A reduction in a river's base flow also decreases the dilution of municipal and industrial wastes added downstream, posing pollution and health hazards. The deterioration of water quality below an irrigation project can render the water unfit for other users, harm aquatic species, and, because of high nutrient content, result in aquatic weed growth that clogs waterways and has health, navigation and ecological consequences.

7. The potential direct negative environmental impacts of the use of groundwater supplies for irrigation arise from overtapping groundwater supplies (withdrawing water in excess of the rate of recharge). This results in the lowering of the water table, land subsidence, decreased water quality, and saltwater intrusion (in coastal areas).

8. A number of external environmental factors influence irrigation projects. Upstream land use will affect the quality of water entering the irrigation area, particularly the sediment content (e.g., from agriculturally-induced erosion) and chemical composition (e.g., from agricultural and industrial pollutants). Use of river waters with a large sediment load may result in canal clogging. Over time, cleaning the canals and depositing the sediment on cropland, or simply irrigating with water of high sediment content can raise the land level to such a height that irrigation is impaired.

9. The obvious benefits conferred by irrigation are those resulting from increased production of food. In addition, concentration and intensification of production on a smaller area can protect forests or wildlands from being converted to agriculture. Increased vegetative cover for a greater portion of the year helps reduce soil erosion, as does land preparation (e.g., land levelling and contouring). Some health benefits result from improved hygiene and a decrease in the incidence of certain diseases. Irrigation projects can also moderate flooding downstream.

Special Issues

Waterlogging and Salinization

10. Waterlogging and salinization of soils are common problems associated with surface irrigation. On a global basis it has been estimated that annually irrigation takes out of production as much land as it puts in because of soil deterioration, principally salinization. Waterlogging results primarily from inadequate drainage and over-irrigation, and to a lesser extent from seepage from canals and ditches. Salinity problems, naturally more acute in arid and semi-arid areas which have more rapid surface evaporation and saline soils, are exacerbated by irrigation. Waterlogging concentrates salts, drawn up from lower in the soil profile, in the plants' rooting zone. Alkalization (the buildup of sodium in soils) is a particularly detrimental form of salinization which is difficult to rectify. While soils in arid and semi-arid zones have a natural tendency toward salinization, many of soil-related problems could be minimized by installing adequate drainage systems. Drainage is a critical element of irrigation projects that too often is poorly planned and managed. Waterlogging and salinization can also be reduced or minimized by using sprinkler or drip irrigation which apply water more precisely and can more easily limit quantities to no more than the crop needs.

Social Issues

11. Social disruption is inevitable in large irrigation projects covering vast areas. Local people dislocated by the irrigation project face the classic resettlement problems: a decrease in the standard of living, increased health problems, social conflicts, and deterioration of natural resources in the resettlement area (see "Involuntary Resettlement" section). The people remaining in the area will likely have to change their land use practices and agricultural patterns. Those moving into the area to benefit from the irrigation scheme similarly will have to adapt to new conditions (see "New Land Settlement" section). The local people often find that they have less access to water, land and vegetation resources as a result of the project. Conflicting demands on the water resources and inequalities in distribution can easily occur both in the project area and downstream. All these factors -- changing agricultural practices, increasing population density, and altering the distribution of wealth -- can have a profound influence on traditional social patterns.

12. An increase, sometimes extraordinary, of water-borne or water-related diseases commonly is associated with the introduction of irrigation. The diseases most often linked with irrigation are schistosomiasis, malaria and onchocerciasis, whose vectors proliferate in the irrigation waters. Other irrigation-related health risks include those associated with increased use of agrochemicals, deterioration of water quality, and increased population pressure in the area (see "Public Health and Safety" section).

13. The reuse of wastewater for irrigation has the potential of transmitting communicable diseases (mainly helminthic, and to a lesser extent bacterial and viral). The population groups at risk include agricultural workers, consumers of produce (and meat) from the wastewater-irrigated fields, and people living nearby. Sprinkler irrigation poses an additional risk through the potential dispersal of pathogens through the air. These risks vary according to the extent of treatment given to the wastewater prior to reuse.

Irrigation Efficiency and Improvement of Existing Systems

14. Inefficient use of water (i.e., overwatering) not only wastes water which could go to other users and avoid ecological impacts downstream, but results in land deterioration through waterlogging, salinization and leaching, and decreased crop productivity. Maximizing the efficiency of water use, therefore, should be of primary concern to all irrigation projects.

15. As already stated, large areas of irrigated land has gone out of production because of soil deterioration. It may be cost effective and certainly beneficial environmentally to invest in land restoration rather than in increasing the area in irrigation.

Project Alternatives

16. A variety of alternatives to a proposed irrigation project, its design, and management exist. They are as follows: (a) improve the efficiency of existing projects and restore degraded crop lands to use rather than establishing a new irrigation project; (b) develop small-scale, individually owned irrigation systems as an alternative to large, publicly owned and managed schemes; (c) develop irrigation systems using groundwater resources which have less potential for causing environmental damage than surface

water systems; (d) develop, where possible, irrigation systems using surface and groundwaters conjointly to increase the flexibility of water supply and minimize negative hydrological impacts; (e) use sprinkler or drip irrigation as alternatives to surface irrigation to decrease the risk of waterlogging, salinization, erosion and inefficient water use; (f) site the irrigation project on the site where negative social and environmental impacts are minimized; and (g) use treated wastewater, where appropriate, to make more water available to other users or reduce the environmental impacts of withdrawing water from surface and groundwater supplies.

Management and Training

17. Institutional factors are often cited as the cause of failure of large scale public irrigation schemes. Operation of all control facilities from the water source to individual farms requires almost constant management. Careful water management is essential to the quantity, timing, controllability, and predictability of water delivered to the users, all of which will determine the success of the project. Training of a cadre of managers to provide the needed services is required if they are not available or lack necessary technical and managerial skills.

18. Planning and implementation of an irrigation project must be done with the cooperation and collaboration of engineers, soil scientists, hydrologists, public health specialists, social scientists and economists. An operations plan, outlining the operating rules and water distribution goals, should be developed prior to the design of the physical infrastructure and guide the subsequent project management.

Monitoring

19. Factors to be monitored should include: climate (wind, temperature, rainfall, etc.); stream discharge above the irrigation project and below at various points; nutrient content of discharge water; flow and water levels at critical points in the irrigation system; water table elevations in project area and downstream; water quality of project inflows and return flows; quality of groundwater in project area; water salinity levels in coastal wells; physical and chemical properties of soil in irrigation area; agricultural acreage in production; cropping intensity; crop yield per unit of land and water; erosion/sedimentation rates in project area; relation between water demand and supply of users (equitability of distribution); condition of distribution and drainage canals (siltation, presence of weeds, condition of linings); upstream watershed management (agricultural extent and practices, industrial activity); incidence of disease and presence of disease vectors; health condition of project populations; changes in natural vegetation in the project area and on the floodplain downstream; changes in wildlife populations in the project area and on the floodplain downstream; and fish population and species.

Table 8.7. Irrigation and Drainage

Potential Negative Impacts	Mitigating Measures
Direct	
1. Soil erosion (furrow, surface).	1. • Proper design and layout of furrows or field avoiding too steep a gradient. • Land leveling. • Design of terraces on hillside minimizing surface erosion hazard.
2. Soil erosion (with sprinkler irrigation on hilly area).	2. Design of sprinkler system minimizing erosion hazard assuring infiltration rate exceeds application rate of the sprinklers.
3. Waterlogging of soils.	3. • Regulation of water application to avoid overwatering (including controlled turn-out to allow cutting off water supply to irrigation ditches). • Installation and maintenance of adequate drainage system. • Use of lined canals or pipes to prevent seepage. • Use of sprinkler or drip irrigation.
4. Salinization of soils.	4. • Measures to avoid waterlogging: • leaching of salts by flushing soils periodically • cultivation of crops with salinity tolerance

Table 8.7. Irrigation and Drainage (continued)

Potential Negative Impacts	Mitigating Measures
Direct (continued)	
5. Scouring of canals.	5. Design of canal system to minimize risk and use of lined canals.
6. Clogging of canals by sediments.	6. • Measures to minimize erosion on fields. • Design and management of canals to minimize sedimentation. • Provision of access to canals for removal of weeds and sediments.
7. Leaching of nutrients from soils.	7. • Avoidance of overwatering. • Replacement of nutrients by fertilizers or crop rotations.
8. Algal blooms and weed proliferation.	8. Reduction of input to and release of nutrients (nitrogen and phosphorous) from fields.
9. Clogging of canals by weeds.	9. • Design and management of canals to minimize weed growth. • Provision of access to canals for treatment or removal of weeds.
10. Deterioration of river water quality below irrigation project and contamination of local ground water (higher salinity, nutrients, agrochemicals) affecting fisheries and downstream users.	10. • Improved water management; improved agricultural practices and control of inputs (particularly biocides and chemical fertilizers). • Imposition of water quality criteria.

Table 8.7. Irrigation and Drainage (continued)

Potential Negative Impacts	Mitigating Measures
Direct (continued)	
11. Sea water intrusion into downstream freshwater systems.	11. • Reduction of takeoff to maintain adequate downstream flow. • Recharge of coastal aquifers through injection wells.
12. Reduction of downstream flows affecting flood plain use, flood plain ecology, riverine and estuarine fisheries, users of water, dilution of pollutants.	12. • Relocation or redesign of project. • Regulation of takeoff to mitigate effects. • Compensatory measures where possible.
13. Encroachment on swamps and other ecologically sensitive areas.	13. Siting of projects to avoid or minimize encroachment on critical areas.
14. Alteration or destruction of wildlife habitat or impediment to movement of wildlife.	14. • Siting of project to minimize loss or avoid encroachment on most sensitive or critical areas. • Establishment of compensatory parks or reserved areas. • Animal rescue and relocation. • Provision of corridors for movement.
15. Impediment to movement of livestock and humans.	15. Provision of passageways.

Table 8.7. Irrigation and Drainage (continued)

Potential Negative Impacts	Mitigating Measures
Direct (continued)	
16. Threat to historic, cultural or aesthetic features.	16. • Siting of project to prevent loss. • Salvage or protection of cultural sites.
17. Alteration or loss of flood plain vegetation and disturbance of coastal ecosystems (e.g., mangroves).	17. • Siting of project to less vulnerable area. • Limitation and regulation of water take-off to minimize problems to extent possible.
18. Dislocation of populations and communities.	18. • Siting of project to minimize effect. • Resettlement scheme ensuring at least equal standard of living.
19. Introduction or increase in incidence of water-borne or water-related disease (schistosomiasis, malaria, onchocerciasis, etc.).	19. • Prevention measures: • use of lined canals or pipes to discourage vectors • avoidance of stagnant or slowly moving water • use of straight or slightly curving canals • installation of gates at canal ends to allow complete flushing • filling or draining of borrow pits along canals and roads • disease prophylaxis • disease treatment

Table 8.7. Irrigation and Drainage (continued)

Potential Negative Impacts	Mitigating Measures
Direct (continued)	
20. Disease and health problems from use of wastewater in irrigation.	20. • Wastewater treatment (e.g., settling ponds) prior to use. • Establishment and enforcement standards for wastewater use.
21. Conflicts over water supply and inequalities in water distribution throughout service area.	21. Means to ensure equitable distribution among users and monitor to assure adherence.
22. Overpumping of groundwater.	22. Limitation of withdrawal so that it does not exceed "safe yield" (recharge rate).
Indirect	
23. Increased pollution and health hazards from downstream industrial and municipal pollutants caused by decreased flow (decreased dilution) of river water.	23. • Control of waste sources downstream. • Reduction of water take-off.
External	
24. Water quality deteriorated or made unusable by upstream land use and pollutants discharge.	24. • Control of land use in watershed areas. • Control of pollution sources. • Water treatment prior to use.

LIVESTOCK AND RANGELAND MANAGEMENT

1. The World Bank finances livestock development projects or project components targeting the improvement of rangeland condition and productivity; the health and productivity of livestock (including cattle, sheep, goats, camels, buffalo, oxen and swine) for meat, milk, hides and fiber; and the welfare of herders. Rangelands include grasslands, open forest (and in some places cleared areas of closed forests), shrublands, and deserts that support domestic ruminants and wild herbivores. The targets of assistance range in scale from large commercial ranching to subsistence mixed-farming or traditional pastoral systems. In both large and small scale operations, the production systems can vary in intensity, with the more intensely managed systems feeding harvested natural or cultivated feed to penned animals, and the more extensive systems relying exclusively on grazing of natural vegetation.

2. The emphasis here will be primarily on small and medium scale livestock operations practicing less intensive management systems (free grazing), with more limited reference to large commercial enterprises (less common in Bank lending) and systems that stall feed animals (which have fewer potential environmental problems than free grazing management systems). Processing of meat and other livestock products is covered in the "Agroindustry" section.

3. Projects involving other uses of rangeland are possible. Wildlife conservation, water catchment, tourism, recreation, hunting, mining and oil exploration are examples.

Potential Environmental Impacts

4. Livestock grazing is a productive land use in areas unsuited to agricultural cropping. Generally practiced in arid or semi-arid lands where rainfall is sparse and unpredictable in time and space, extensive livestock production, particularly pastoralism, is an appropriate and sustainable form of land use that is much less risky than cropping. Grazing and rangeland vegetation are coadapted; both undergrazing and overgrazing, resulting in the growth of woody plants and large, unpalatable grasses, reduce the productive potential of an area. Grazing also helps maintain soil fertility and physical soil characteristics by deposition of animal dung. And, the germination of certain plants is enhanced or made possible when seed is passed through the digestive tract of animals. Livestock production, therefore, represents a system of land management in marginal areas that can maximize food production with minimal input while maintaining the productivity of the ecosystem.

5. Negative environmental impacts of livestock grazing, however, arise from overgrazing and result from various range management practices. External impacts on the rangelands include development activities (e.g., agriculture, water resource development, settlement programs, mining, etc.) which reduce or eliminate grazing on the land or degrade range resources.

6. The major potential negative environmental impact of livestock production is overgrazing or overuse (e.g., excess harvesting) of fodder and forage resources leading to degradation of the vegetation, increased soil erosion, and deterioration of soil fertility and soil structure (see Table 8.8 at the end of this section). Overgrazing is a consequence of overstocking the land so that the number and type of animals exceed the area's carrying capacity. Overgrazing results in the decrease of favored fodder species and

the increase of unpalatable, weedy species. It increases soil erosion, indirectly by decreasing vegetative cover, and directly by loosening the soil and making it subject to water and wind erosion. Livestock paths scar hillsides and trigger an erosional process that may result in gully formation. Grazing also degrades the structure of the soil by pulverizing it and compacting the surface, thus lowering the soil's infiltration capacity. Rangeland management techniques to reduce grazing pressure include altering the timing, length or succession of use by livestock on particular land areas, and regulating the numbers, species and movement of animals. The most common ways of controlling animal pressure are through the use of rotational grazing, deferred grazing and strategic placement of water points and salt. Organized marketing of range products and development of dry-season grazing areas and grazing reserves are the other useful management techniques.

7. Increasing livestock production on rangelands or introducing animal husbandry into areas may have negative impacts on wildlife populations. Competition for vegetation and water resources may increase, and livestock owners may kill wildlife for food or to eliminate them as pests (i.e., predators on livestock species). Livestock and wildlife have coexisted successfully in many places, utilizing different resources and thus avoiding excessive competition. In some areas wildlife ranching has shown greater potential than livestock production, and may be considered as an alternative for meat, hides and skin production. Tourism and controlled hunting are other alternatives to consider.

8. Vegetation depletion and increased soil erosion often occur around water points where animals congregate. Increasing the number of water points, locating them strategically, and closing them at certain times of the year can limit resource destruction. Poor planing, placement, management and control of water holes, however, can compound problems and aggravate the severity of droughts. The drilling of deep wells can also result in severe land degradation around the boreholes by lowering the watertable and thus affecting local vegetation. The dual use of water points for livestock watering and human water supply has obvious negative health implications.

9. Rangeland management techniques to increase the productivity of the rangeland include: mechanical and physical work on the soil or vegetation (e.g., land contouring and other soil and water conservation techniques, bush clearing); planting, seeding or reseeding with selected species and varieties; burning the vegetation; fertilization with manure or chemicals and pest control efforts. Soil and water conservation measures and seeding of vegetation can decrease soil erosion, while bush clearing and burning, if not carefully carried out may increase erosion. Not only does site productivity decline with increased erosion, but water bodies suffer from increased levels of sediment input.

10. Burning is the oldest practice used by man to manipulate range vegetation for livestock grazing. Burning is done to control undesirable brush and tall weeds, destroy old, unpalatable grass tussocks, and favor the growth of fresh, more digestible and nutritious grasses. Fire increases forage yield and palatability of grasses and forbs. Haphazard or accidental burning, however, can be harmful or disastrous for the vegetation and soils, and can lead to increased levels of soil erosion.

11. Use of chemicals for fertilization of pasture or for pest or disease control can have negative environmental impacts. Chemical fertilization is prohibitively expensive and thus rarely practiced in most developing countries. Where used, however, it can cause water pollution problems, as can fertilization with organic materials. Use of chemicals as herbicides or for disease control (e.g., for reduction of the tsetse fly population to control trypanosomiasis) is more common and can have negative effects on wildlife, water supplies (surface and ground waters) and vegetation.

12. Livestock improvement includes improved veterinary care, disease treatment and control, and breeding or other stock improvement techniques. Increases in livestock populations resulting from these efforts must be done in conjunction with range management and grazing control in order to avoid problems from increased pressure on the range resource leading to overgrazing. Selective breeding has the long-term negative potential of decreasing the natural genetic variability of livestock populations and thus reducing their resistance to disease and flexibility to adjust to climatic change.

13. Some projects provide supplemental feeding assistance during times of drought to maintain livestock herds. Care must be taken with such subsidized feeding programs that the food assistance is continued until range resources have adequately recovered from the drought. A misconception is that once the rains start again livestock feeding programs can be discontinued, but, in fact, there is a lag time between the time the rains start and the range is ready to accommodate high grazing pressure again. A great deal of damage can be done to grasslands by turning the livestock out too early.

Special Issues

Livestock Production in Moist Tropical Forests

14. Clearing of moist tropical lowland forest lands or using land already cleared for other reasons for livestock production has proven to be an unsustainable practice with disastrous environmental impacts. Clearing of large areas of forest land in Amazonia and subsequent conversion to cattle production, has degraded the land irreparably and rendered it useless for other purposes. Large-scale cattle ranching in these areas is highly unsuitable and should be discouraged.

Social and Cultural Issues

15. Rangeland management systems and socio-economic patterns and conditions are intimately linked. A decline in rangeland productivity will through natural (e.g., climatic) or humanly driven forces, will have negative impacts on family income, health, and the distribution of scarce resources between people. Conversely, socio-economic factors such as labor availability, distribution of tasks within families, land use and resource rights, property ownership patterns, and market conditions affect how the range and livestock resources are managed.

16. Many rangeland areas are in a state of transition. The most important social and economic changes in these areas include: (a) increased involvement in wage-labor markets; (b) transformations of indigenous tenure systems and organizations; (c) encroachment of dryland agriculture and irrigation into rangelands; (d) increased involvement of pastoralists in commodity markets; (e) increased sedentarization and settlement that is often encouraged by state and donor policies and programs; and (f) fluctuating terms of trade for pastoral produce. In terms of their potential effects on the physical environment, the most important variables to identify are income and welfare levels, labor availability, and land/population ratios. Changes in these are likely to affect how physical resources are managed, as are changes in people's traditional access to resources.

17. The key social issues which must be considered before developing any livestock or rangeland management project are spacial and temporal dimensions of rangeland economics; resource tenure and local management; property rights, distribution, and welfare; and labor availability.

18. Livestock production systems in rangeland areas are land extensive. The mobility of human populations in rangelands tends to increase in relation to aridity, with the most mobile populations residing in arid and semi-desert zones. It is critical, therefore, to identify the physical expanse and ecosystem types that livestock exploit over a full "drought-to-drought" cycle (7-10 year period), particularly the lands and water resources that are critical to the survival of the overall production system. Loss of access to lands due to agricultural encroachment (either by outsiders or the herders themselves), settlement projects, the creation of national parks, and other activities will reduce land available to extensive rangeland production and decrease land/population ratios. Critical dry-season areas diverted to non-range use can be particularly detrimental to the livestock herders. Finally, reduced mobility due to physical barriers (e.g., fences, settlements) or government policies favoring sedentarization and settlement will affect the flexibility necessary for optimizing the use of the range resource.

19. In areas where rangeland is a common property resource, most local communities have (or had) either formal or informal institutions for regulating access to pasture, water and other resources. The use of communal land was often based on land ownership or user rights over various ecological zones with nomadic production systems. Government policies to settle nomads, the desire of the population to become sedentary, or appropriation of land for irrigation schemes or other development projects have reduced the total land area and ecological distribution to the point of radically changing social structures and production systems. These changes either preclude livestock production or require increased imports for animals. Government policies on subsidies, taxes and product price fixing have often favored crops over livestock with the resulting misuse of lands with little or no economical crop production potential. This has further exacerbated the impact on social and cultural systems. It is critical to identify both these traditional local rules for regulating access to resources (and how these rules have changed over time) and the government policies and programs influencing local tenure practice. The local organizations and individuals responsible for overseeing tenure regulations, and the class, age and gender characteristics of these organizations, should be examined. Not only should these entities be involved in the design of a project, but equal consideration should be given to those which will be affected, particularly when their concerns are not adequately represented by the controlling bodies. The pros and cons of altering property ownership patterns (e.g., shifting from a system of common property resources to private land ownership) must be carefully weighed in light of local social patterns.

20. It is essential to understand existing local livestock ownership patterns in designing a rangeland management project. Livestock ownership in most rangelands is highly skewed, often with 10 percent of herders controlling up to 50 percent of livestock units with large segments (frequently 20 to 30 percent of households) owning little or no livestock or other property. It is dangerous to try to adjust livestock numbers to correspond to the carrying capacity of the range without first understanding property ownership dynamics and the number of animals per household necessary to support the household and maintain an economically-viable system. The danger is that if the number of animals per household is limited (rather than the number of household units involved in livestock production) to reduce grazing pressure, the number may be reduced to an uneconomic management unit, thus forcing the poorest households out of production.

21. Income distribution and intrahousehold property rights (e.g., those based on gender and age) are also frequently skewed. In addition to information on the general distribution of property and income among and within households, the following factors must be understood: the extent to which management and ownership of property are vested in the same unit (e.g., wealthy segments of the populations, from either within or outside the community, acting as absentee owners, and hiring local herders to tend their animals); the extent to which certain categories of property (e.g., goats and sheep) or sources of income (e.g., income from milk sales) are controlled by women and if development activities have eroded the income-earning activities of women; and the extent to which certain types of resource management and production activities can be distinguished by wealth or income category.

22. Since rangeland production systems generally are carried out in areas of low population density, they often suffer from labor shortages. Lack of labor affects herd movement, mobility, and certain conservation and resource management techniques. The issues to be assessed include: diversification of the local production system (the degree that mixed cropping, wage labor opportunities and other non-farm activities affect the availability of labor for livestock production); the extent to which engagement in nonfarm activities by males increases the workload of women; and the extent to which labor shortages and low incomes affect local management strategies and are associated with disruptive resource management practices.

23. In areas which are more densely populated, with limited range resources, or in ecologically sensitive areas such as mountainous lands, restriction of livestock movement may be encouraged to limit environmental degradation by shifting to a system of stall feeding animals. While this may be environmentally more sound, access rights to forage and fodder resources may be unequitable, or the concentration of ownership of land on which livestock feed can be grown may preclude the involvement of the poorer and landless in such a scheme. Such factors must be taken into consideration in project design.

Project Alternatives

24. There are few alternatives to livestock production in rangelands as it is generally the most suitable land use available for the climatic conditions. One possibility, although limited to certain areas and circumstances, is wildlife ranching. Management of wildlife as a sustainable system potentially can increase the productivity of the land in terms of meat, hides, skins and other products while minimizing environmental destruction. This success of such a system, however, depends on numerous variables, not the least of which is marketing. Wildlife-based tourism, recreation, controlled hunting, and rainfed dryland farming are other alternatives.

25. The greatest scope for alternatives lies within the livestock management system itself. As already discussed, choices in variables include the following: species of animal(s) for production; intensity of production; rangeland improvement activities; sprinkler irrigation; livestock improvement activities; transport and processing arrangements; and control of marketing functions.

Management and Training

26. Government policies and programs designed to increase livestock production often affect the range resource and welfare of producers negatively and promote economic inefficiencies and social inequities. Livestock policies have often worked counter to development assistance in livestock production. Policies, programs and legislation should be reviewed for their social and environmental impacts. Producer groups which can provide services, represent the individual producers, and facilitate marketing may need to be established or strengthened.

27. Government policies (e.g., on livestock prices and import and export controls) are often aimed at keeping consumer prices for livestock products low. Then to compensate the producers, governments offer subsidies allowing producers to purchase inputs below their real value, provide free services or offer subsidized credit. This creates an artificially bolstered system which may lead to inefficiencies, inequities (favoring the large over the small producers), investment distortions, and the degradation of range resources through inappropriate land use. It also tends to favor capital intensive operations using feed concentrates over forage-based systems.

28. In many places government services are inadequate. In others, some services are excessive and stifle private sector initiatives which could provide the services more efficiently. The governments of most countries are involved in livestock processing, marketing, supplying of inputs, extension of technical services, and programs in animal health and genetic improvement. The delivery of these services is often poor.

29. Government research and extension services are often lacking. They commonly focus on intensive, import-dependent production systems which are not financially viable for the majority of livestock producers. Extension services are often hampered by inefficiencies, lack of funds, and inadequately trained extension agents.

30. Traditional social structures are breaking down and are not being replaced by producer organizations to monitor use of common lands. Legislation defining land ownership and user rights, particularly on common lands, is often weak or unenforced. The lack of producer organizations hinder marketing and the gap is filled by middlemen who siphon off the majority of profits. Producer groups can establish and regulate land use patterns, livestock management systems and livestock numbers; provide services now provided by the government; and represent the livestock producers when working to influence government policies and research priorities.

31. Training is often needed to bolster government personnel's technical, managerial and extension skills. Common training needs are in range ecology and management, livestock production systems, veterinary sciences, agricultural economics, extension techniques, research and management skills.

32. Research should be tailored to producers' needs, particularly in pasture and range production. Research is needed on developing technological packages suitable for the small-scale producers with low labor input. Other research needs include that on farm forestry which integrates livestock and agricultural cropping with tree growing patterns, and practices of pastoralists to investigate their ecological and economic viabilities.

Monitoring

33. Factors to be monitored in a livestock project should include:

- range condition (assessment of present state of health of the range in relation to its potential)
- range trend (direction of change of range condition)
- availability of and access to natural fodder and forage, cultivated fodder, and imported feedstuffs (in stallfed animals)
- numbers and types of animals
- seasonal distribution and movement of animals
- condition of the livestock (weight, presence of disease, other health indices)
- condition of the soil (i.e., signs of increased erosion, compaction, decreased fertility, etc.)
- water points (location, condition, and intensity of use, and condition of vegetation around the water points)
- market conditions (changes in price, development of alternative markets, etc.)
- changes in economic indices of livestock producers (e.g., income levels and health)
- changes in social organization
- external land use changes and demographic changes which have impacts on the range resource and livestock producers
- changes in wildlife populations and habitat due to livestock production

Table 8.8. Livestock and Rangeland Management

Potential Negative Impacts	Mitigating Measures
Direct	
1. Degradation of vegetation resources due to overgrazing.	1. • Limitation of animal numbers. • Control of length of grazing time on particular areas. • Mixing of livestock species to maximize use of vegetation resource. • Reseeding and fodder production. • Cut-and-carry. • Strategic placement of water points and salt.
2. • Increased soil erosion due to clearing of vegetation and trampling. • Increased saltation of surface waters.	2. • Restriction of livestock access to unstable areas (e.g., stoop slopes). • Soil erosion control measures (e.g., reforestation, reseeding of grasses, land preparation, terracing).
3. • Deterioration of soil fertility and physical characteristics through: • removal of vegetation • increased erosion • soil compaction	3. Same as 1 and 2.
4. Increased rapid runoff due to vegetation clearing and soil compaction (decreased infiltration capacity).	4. • Water conservation measures and water spreading. • Same as 1 and 2.

Table 8.8. Livestock and Rangeland Management (continued)

Potential Negative Impacts	Mitigating Measures
Direct (continued)	
5. • Degradation of vegetation and soil around water points. • Overtapping of groundwater. • Lowering of watertable and degradation of vegetation locally by drilling and use of boreholes.	5. • Development of many small-capacity water sources. • Strategic placement of water points. • Control of use of water points (animal numbers and time of year). • Closure of permanent water sources when temporary pools and streams are available. • Limitation of well capacity by choice of technologies (e.g., handpumps or buckets instead of motor pumps).
6. • Displacement or reduction of wildlife populations by reduction of habitat. • Disruption of migratory routes. • Competition for food and water resources. • Introduction of diseases. • Impacts of burning. • Increased poaching and killing of wildlife considered as pests or predators to livestock.	6. • Planning and implementation of range management strategies (choice of species, livestock numbers, grazing areas) that minimize negative impacts on wildlife. • Establishment of compensatory wildlife refugees. • Investigation of management of wildlife ranching which will help protect wildlife resources.

Table 8.8. Livestock and Rangeland Management (continued)

Potential Negative Impacts	Mitigating Measures
Direct (continued)	
7. Pollution environmental disruption and health hazards from disease and pest control measures.	7. • Choice of chemical that is species-specific, short residence time (active period), and has low impact on other biologic resources. • Protective measures for field workers. • Spraying methods and timing to minimize potential of water pollution. • Selection of disease-resistant livestock breed.
8. Reduction of genetic variability due to selective breeding.	8. Conservation of genetic diversity in-site (protection of wild relatives in natural habitat, maintaining variability within populations by breeding) and ex-situ (e.g., preservation of genetic material in "banks").
9. Negative effects of uncontrolled burning for brush control on soil and vegetation (deterioration of soil fertility and soil structure, altered wildlife habitat, destruction of vegetation).	9. Implementation of well-planned and controlled burning programs.
10. Conversion of moist tropical lowland forests for livestock production resulting in long-term environmental degradation and unsustainable production.	10. Avoidance of clearing such forests for livestock production.

RURAL ROADS

1. Rural roads connect villages and smaller population centers such as regional market towns, and comprise tertiary, feeder and penetration roads. They usually are unpaved or only lightly paved, are narrower and have tighter curves and steeper grades than highways. They may be all-weather or only seasonal and often include fords or ferries rather than bridges. Highways, whether paved or unpaved, which traverse rural regions are treated in the "Roads and Highways" section of Chapter 9.

2. Rural road construction, improvement or rehabilitation can be undertaken as a transportation sector loan or as a subcomponent of another project such as in agriculture, forestry or rural development. When they are subcomponents, they may be difficult to detect in project documents and even the borrower may not know where they will be geographically. Projects which develop location-specific resources far from existing roads (e.g., mineral deposits, timber resources, hydropower installations, or tourism sites) will entail construction of access routes. These roads may have to be paved when the transport of heavy loads is involved.

Potential Environmental Impacts

3. Environmental impacts include: direct impacts at the road construction site and in the immediate environs of the right of way and indirect impacts in the adjoining region. This larger area of influence of a rural road is the area of induced economic, social or environmental effects, whether planned or spontaneous, which are consequences of increased physical access and lower transportation costs. These indirect impacts often greatly exceeds the direct impacts.

4. Air and water pollution and noise, generally associated with highways, are not major problems in rural roads because traffic density is low. Dust raised from the road and blown by vehicles can pose a health hazard and damage vegetation along the side of the road. If the road is paved, some water pollution can occur from chemicals carried away in runoff. Other possible sources of water pollution are chemicals applied along the roadside or on the right-of-way for the control of weeds or dust.

5. The major direct environmental impact associated with rural roads is erosion. Construction in the rainy season, or improper construction methods which leave soils exposed unnecessarily, can cause significant erosion. Improper drainage from roads in areas of high precipitation can ruin roads and have adverse impacts on adjacent lands, most pronounced in steep areas. Even in low rainfall areas, if the drainage system is improperly designed, a short but intense rain can erode large swathes downslope of the drainage ditches. Large-scale destruction also can occur through slips and landslides in steep areas. Damaging to land and vegetation resources, erosion also causes serious sedimentation and siltation problems in nearby surface waters. With their greater potential for causing environmental problems, roads built in steep, humid zones will require higher standards and costs than roads on plains. (For further discussion, see Table 8.9 at the end of this section.)

6. The design of roadways can also determine the effect of the road on its surroundings, for example, in large wetland areas (e.g., tidal flats, lagoons or inland deltas). Roads built on raised embankments which interfere with cross drainage and on causeways which do not adequately allow for

movement of water can permanently impair the biological cycles and productivity of the wetland ecosystems. Similarly, roads can cause flooding of adjacent areas by blocking the flow of water and by increasing runoff.

7. Some roads constructed by private industries (e.g., for mining, ranching, logging) may be later formalized and converted to public roads, due to local petitioning and pressure on ministries of public works. The routes of such roads, however, may not serve public needs or be compatible with land use capabilities. Road rehabilitation projects may have to be launched to remedy the problems of such ill planned roads. Some logging roads built to gain access to steep lands in humid regions (e.g., Malaysia) have permanently damaged relatively large portions of such lands.

8. The construction of penetration roads into remote areas triggers migration into adjacent lands and induces land uses and resource exploitation patterns that are extremely difficult to manage or control (see sections on "Induced Development" and "New Land Settlement" in Chapter 3). Roads intended to open up moist tropical forest lands to settlement, for example, in Kalimantan or the Transamazonica in the Amazon Basin, have almost invariably triggered spontaneous migration into the area and resulted in a sequence of uncontrollable events characterized by wasteful logging methods, invasion of tribal lands and conversion of land to farming of short cycle crops which is unsustainable. Outcomes have included severe disruption of indigenous peoples, conversion of forest to pasture land, invasion of weedy species and land degradation resulting in eventual abandonment of the area. Preventing or mitigating unwanted changes where there are no alternatives to road construction may require a long-term companion development effort.

Special Issues

Road Siting

9. Siting of a road is the most critical decision in road construction. It will largely determine the type and magnitude of environmental and social impacts that will result from road construction. Alignments through lands of indigenous peoples, critical wildlands and wildlife habitat, lands unsuited to probable land use changes that will occur by both planned and unplanned development stimulated by the road, and locations where there are potential natural hazards, should be avoided. Once the area for the road is chosen, care must be taken to avoid constructing it on steep slopes, unstable soils and places where runoff and drainage will be problematic. Decisions on road siting frequently involve rapid screening and appraisal of many proposed locations and road specifications, and should involve a wide range of line agencies and levels of organization. Siting involves the collection of data on the climate, soil, geology, topography, hydrology, biology/ecology, and social factors (land and resource use patterns, local economy, class and economic structure, local administrative or power structures) of the sites proposed for construction. Remote sensing information sources and qualified interpreters are particularly useful for these functions.

Road Maintenance

10. Designing rural roads projects poses several special problems. Maintenance costs are habitually underestimated and maintenance routines deficient, especially on low volume roads (e.g., roads through lightly populated regions, extensive ranching areas or widely separated settlements). Adequate road maintenance can be costly but is essential to avoid environmental and socio-economic problems.

Overloading of freight vehicles is a chronic problem that should be taken into account in road design and maintenance. To the extent possible, road maintenance should be supported technically and financially (at least at first) by the central government, but carried out by local authorities and the community.

Social Impacts

11. The construction of a rural road can confer a host of benefits on the local populations. These include: improved access to markets; increased government services, such as electricity, potable water, extension services, credit systems, health services and education; stimulation of agroindustries; and greater employment opportunities, at least in the short term. Although all positive contributions to rural development, these benefits often are not evenly distributed among population groups and may exaggerate socio-economic disparities. The larger, wealthier landowners generally have better access to services than the poorer peoples.

12. The introduction of rural roads leads to increased land values and more intensive land use, especially for land adjacent to the road. The result is often increased rents, or shifts in ownership or rights of resource use from the poorer to wealthier classes. Land located back from the road may decline in value.

13. Changes in agriculture are often characterized by intensification of production and shifts from subsistence to cash crops. Dependence on cash crops to the exclusion of subsistence crops can have negative effects on local nutrition levels. Roads, by encouraging mobility of people and goods, can also facilitate the spread of pests and disease.

14. Unempowered ethnic minorities, who previously have lived in geographic and political isolation from the rest of the country, often receive little or no benefit from penetration roads. In fact, the construction of a road into their lands can be devastating to these people. In some areas, for example Amazonia, roads have resulted in dispossession of indigenous peoples' lands, malnutrition, decimation of their numbers through the introduction of diseases, cultural disintegration, and a host of other social and economic problems. (See "Indigenous Peoples" section in Chapter 3.)

Project Alternatives

15. There are few alternatives that fulfill the same functions as rural roads. Water transport is a viable alternative in regions with navigable rivers and streams, but such proposals frequently have been rejected. Narrow gauge railways are alternatives for transporting mineral ores. Under some circumstances, air transport may be a viable alternative when all the financial, social and environmental costs of the road are calculated.

Management and Training

16. The planning, construction and maintenance of a road involves a variety of line agencies and levels of organization (national, regional and local), including the central government, planning and financing ministries, one or more operating ministries, the executing unit, private contractors, nongovernmental agencies and local community groups. Input from a variety of agencies will be

necessary at the planning stage, and should begin at the point when road investments are first considered and their contribution to rural development discussed. Coordination in project preparation may consist of agency cross-checking of location of the road and construction facilities, setting up an interagency planning committee or coordinating council, or incorporating agency criteria and/or staff views in the pre-screening process.

17. Vertical linkages must also be well developed, ensuring that the project is linked equally well to its sources of political and financial support in the central government, and to the local communities to be affected by the road. It is essential to incorporate these communities in the planning process. Building and maintaining rural roads involve the local communities and populations as much as possible, with the eventual goal that they will be able to assume primary responsibility for the road. A poor track record in turning road maintenance over to rural communities is often attributed to the top-down approach of most public works departments and limited public participation in the planning and construction of the roads.

18. The executing agency may specify in contracts to the construction firm mitigating actions and supervisory/monitoring actions to be taken. Environmental protection and monitoring offices may be located within transportation ministries but are more likely to be located in another ministry, e.g., an environmental protection bureau. Personnel may lack experience; equipment, vehicles, and operational budgets may be inadequate; and standards may be weak or vague. It may be necessary to develop pragmatically defined standards, such as for erosion-control measures, and to create a special capability and capacity to plan and execute a monitoring plan.

19. Training may be required for all levels of personnel to ensure that road construction is carried out properly and with the least possible negative environmental and social impacts. Training may be targeted at the following people: road construction engineers and technicians in road construction design technology and methods; crew chiefs and site supervisors in construction techniques; unskilled local labor in road construction; and local agency personnel, local authorities and/or representatives of community organizations who will be responsible for road maintenance.

Monitoring of Construction Impacts

20. Monitoring of the impacts of road construction will consist of looking at the following: (a) the "performance" of the installation after construction; (b) erosion during and after construction; (c) the installation of erosion control and drainage works to ensure that it is adequately done; and (d) verification that proper waste disposal at the construction is carried out (cut and fill material, used oil, human waste, trash, debris, etc.).

21. A monitoring plan should be drawn up, on the basis of chosen mitigating measures and assignment of responsibilities for their implementation. The monitoring of immediate and longer term indirect impacts and their mitigation will fall within the jurisdiction of regional development authorities and ministries, (e.g., agriculture, forestry) according to the nature and scope of development that is induced by, or planned in parallel with, rural road construction. If planned and integral to the development loan, the corresponding sections in the Sourcebook will indicate the kind of monitoring needed. If not integral to a road construction loan, certain impacts may be sufficiently important to justify a companion planning and development exercise, with its own monitoring functions.

Table 8.9. Rural Roads

Potential Negative Impacts	Mitigating Measures
Direct: During Construction	
1. Erosion from fresh road cuts and fills and temporary sedimentation of natural drainage ways.	1. • Limitation of earth moving to dry periods. • Protection of most susceptible soil surfaces with mulch. • Protection of drainage channels with berms, straw or fabric barriers. • Installation of sedimentation basins, seeding or planting of erodible surfaces as soon as possible.
2. Ground and water contamination by oil, grease, and fuel in equipment yards.	2. • Collection and recycling of lubricants. • Precautions to avoid accidental spills.
3. Creation of stagnant water bodies in borrow pits, quarries, etc. suited to mosquito breeding and other disease vectors.	3. Assessment of vector ecology in work areas and employment of measures (e.g., improved landscaping, filling or drainage) to avoid creating habitats.
4. Environmental and social disruption by construction camps.	4. Careful siting, construction and management of construction camps.
Direct: Permanent	
5. Destruction of buildings, vegetation and soil in the right of-way, borrow pit sites, waste dumps, and equipment yards.	5. • Alternative alignments. • Harvest and utilization of public domain forest resources prior to construction.

Table 8.9. Rural Roads (continued)

Potential Negative Impacts	Mitigating Measures
Direct: Permanent (continued)	
	• Compensation given to private landowners. • Restoration of sites to original condition to extent possible through reclamation measures.
6. Interruption of subsoil and overland drainage patterns (in areas of cuts and fills).	6. Installation of adequate drainage works.
7. Landslides, slumps, slips and other mass movements in road cuts.	7. • Route alignment to avoid inherently unstable areas. • Design of drainage works to minimize changes in surface flows and adequate to local conditions, according to prior surveys. • Stabilization of road cuts with structures (concrete walls, dry wall masonry, gabions, etc.).
8. Erosion of lands below the roadbed receiving concentrated outflow carried by covered or open drains.	8. • Increase in number of drain outlets. • Drain outlets placed so as to avoid cascade effect. • Lining of receiving surface with stones, concrete.
9. Increased suspended sediment in streams affected by road cut erosion, decline in water quality and increased sedimentation downstream.	9. • Establishment of vegetative cover on erodible surfaces as soon as possible. • Establishment of retention ponds to reduce sediment load before water enters stream.

Table 8.9. Rural Roads (continued)

Potential Negative Impacts	Mitigating Measures
Direct: Permanent (continued)	
10. Marred landscape (scars from road cuts, induced landslides and slumps, etc.).	10. • Tourist site access roads planned with regard for visual aesthetics. • Grade limitations to avoid cutting and filling where scenery would be spoiled. • Maintenance and/or restoration of roadside vegetation.
11. Health hazards and interference of plant growth adjacent to road by dust raised and blown by vehicles.	11. Dust control by application of water or chemicals.
12. Contamination of ground and surface waters by herbicides for vegetation control and chemicals (e.g., calcium chloride) for dust control.	12. • Reduction of use. • Alternative (non-chemical) methods of control.
13. Accident risks associated with vehicular traffic and transport, that may result in spills of toxic materials (see "Hazardous Materials Management" section), injuries or loss of life (see "Public Health and Safety" section).	13. • Regulation of transport of toxic materials to minimize danger. • Prohibition of toxic waste transport through ecologically sensitive area.
14. Creation of a new pathway for disease vectors affecting humans and animals.	14. Establishment of plant and animal sanitation service and related checkpoints.
15. Disruption/destruction of wildlife through interruption of migratory routes, disturbance of wildlife habitats, and noise related problems.	15. Siting to minimize impacts.

Table 8.9. Rural Roads (continued)

Potential Negative Impacts	Mitigating Measures
Indirect	
16. Unplanned or illegal timber cutting.	16. See "Natural Forest Management" section.
17. Unplanned or illegal land clearing.	17. See "New Land Settlement" section.
18. Long-term or semi-permanent destruction of soils in cleared areas not suited for agriculture.	18. See 17 above.
19. Planned development and illegal invasion of homelands of indigenous peoples by squatters and poachers causing serious social and economic disruption.	19. See "Indigenous Peoples" section.
20. Destruction or damage of terrestrial wildlife habitats, biological resources or ecosystems that should be preserved by induced development.	20. See "Biological Diversity" and "Wildlands" sections.
21. Damaging alteration of wetland ecosystems traversed by causeways.	21. See "Wetlands" section.
22. Excessive and/or destructive development of coastal areas or other use of coral for cement and landfill, destroying parts of reef uniquely endowed recreational environments made accessible by roads.	22. See "Tourism Development" section.

References

Agriculture and Rural Development

Alheritiere, D. 1982. Environmental Impact Assessment and Agricultural Development: A Comparative Law Review. Rome, Italy: FAO.

Food and Agriculture Organization. 1989. The State of Food and Agriculture. Rome, Italy.

Goodland, R. and others. 1985. Environmental Management in Tropical Agriculture. Boulder, Colorado: Westview Press.

____. 1991. "Tropical Deforestation: Solutions, Ethics and Religions." Environment Department Working Paper 43. World Bank, Washington, D.C.

National Research Council Board on Agriculture. 1989. Alternative Agriculture. Washington, D.C.: National Academy Press.

World Bank. 1988. Rural Development: World Bank Experience, 1965-86. Operations Evaluation Department. Washington, D.C.: World Bank.

Integrated Pest Management and Use of Agrochemicals

Ahmad, Y. J. 1982. Pesticide Use on Industrial Crops. Nairobi, Kenya: UNEP.

Asian Development Bank. 1987. Handbook on the Use of Pesticides in the Asia-Pacific Region. Manila, Philippines.

United States Agency for International Development. 1988. Locust and Grasshopper Control in Africa: A Programmatic Environmental Assessment. College Park, Maryland.

World Health Organization. 1986. WHO Recommended Classification of Pesticides by Hazard. Geneva, Switzerland.

World Bank. 1991. "Agricultural Pest Management" (in preparation). Operational Directive 4.02. World Bank, Washington, D.C.

____. 1988. Environmental Guidelines. Environment Department. Washington, D.C.: World Bank.

____. 1987. "Guidelines for the Use, Selection and Specification of Pesticides in Public Health Programs." Operational Policy Note 11.01. World Bank, Washington, D.C.

Agroindustry

Austin, J. E. 1981. Agroindustrial Project Analysis. Baltimore, Maryland: The John Hopkins University Press.

United Nations Economic and Social Commission for Asia and the Pacific. 1982. Industrial Pollution Control Guidelines: IV. Palm Oil Industry. Bangkok, Thailand.

United Nations Environment Programme. 1982. Environmental Aspects of the Sugar Industry--An Overview. Paris, France: Industry and Environment Office.

___. 1982. Hides and Skins Industry. Environmental Operational Guidelines Series. Nairobi, Kenya.

___. 1977. Residue Utilization Management of Agricultural and Agro-Industrial Residues--An Overview. Paris, France: Industry Programme.

World Bank. 1988. Environmental Guidelines: Crumb Rubber, Palm Oil, Leather Tanning and Finishing, Slaughterhouses, Tea and Coffee Production, Wool Scouring. Environment Department. World Bank, Washington, D.C.

Dams and Reservoirs

Cernea, M. M. 1988. Involuntary Resettlement in Development Projects: Policy Guidelines in World Bank-Financed Projects. World Bank Technical Paper 80. Washington, D.C.: World Bank.

Dixon, J. A. and others. 1989. Dams and the Environment: Considerations in World Bank Projects. Washington, D.C.: World Bank.

Goldsmith, E., and N. Hildgard. 1984. The Social and Environmental Effects of Large Dams. 3 Volumes. Camelford, United Kingdom: Wadebridge Ecological Centre.

Goodland, R. 1989. "The World Bank's New Policy on the Environmental Aspects of Dam and Reservoir Projects." Indian Journal Public Administration 25(3):607-633.

Le Moigne, G. J-M and others. 1990. Dam Safety and the Environment. World Bank Technical Paper 115. Washington, D.C.: World Bank.

Mahmood, K. 1987. Reservoir Sedimentation. World Bank Technical Paper 71. Washington, D.C.: World Bank.

World Bank. 1989. "Environmental Assessment." Operational Directive 4.00, Annex A. World Bank, Washington, D.C.

____. 1989. Environmental Guidelines for Dam and Reservoir Projects. Asia Technical Department, Environment Division. Washington, D.C.: World Bank.

____. 1989. "Environmental Policy for Dam and Reservoir Projects." Operational Directive 4.00, Annex B. World Bank, Washington, D.C.

Fisheries

Asian Development Bank. 1987. Checklist for Assessing the Socio-Cultural Impacts of Artisanal Fisheries Projects. Manila, Philippines: Environmental Unit.

Carpenter, R. A., and J. E. Maragos. 1989. How to Assess Environmental Impacts on Tropical Islands and Coastal Areas. Honolulu, Hawaii: East-West Center for Policy and Environment.

European Inland Fisheries Advisory Committee. 1982. "Sessions on Introductions and Transplantations of Exotics; Country Reviews; and Ecological and Practical Considerations." Report of the Symposium on Stock Enhancement in the Mangement of Freshwater Fisheries. Rome, Italy: FAO.

Mahmood, N. 1987. "Effects of Shrimp Farming and Other Impacts on Mangroves of Bangladesh." Workshop on Strategies for the Management of Fisheries and Aquaculture in Mangrove Ecosystems, Bangkok, Thailand, June 1986. FAO Fisheries Report 370 Supplement, FIRI/R370. Rome, Italy.

Petr, T. 1984. "Resources Development Planning and Management and Man-made Impacts on Inland Fish and Fisheries." IPFC Workshop on Inland Fisheries for Planners, August 2-6, 1982, Manila, Philippines. FAO Fisheries Report 288, FIR/R288. Rome, Italy.

Phelps, R. P. 1981. Environmental Considerations in Aquaculture. Alabama: Auburn University, International Center for Aquaculture.

Snedaker, S. C., and C. D. Getter. 1985. Coastal Resources Management Guidelines. Renewable Resources Information Series, Coastal Management Publication 2. Columbia, South Carolina: Research Planning Institute.

Flood Protection

Brookes, A. 1988. Channelized Rivers: Perspectives for Environmental Management. Chichester, United Kingdom: John Wiley and Sons.

Drijver, C. A., and M. Marchand. 1985. Taming the Floods: Environmental Aspects of Floodplain Development in Africa. The Netherlands: State University of Leiden, Center for Environmental Studies.

Frami, K. K., ed. 1983. Manual of Flood Control Methods and Practices. New Delhi, India: International Commission on Irrigation and Drainage.

United Nations Department of Economic and Social Commission for Asia and the Pacific. 1976. Guidelines for Flood Loss Prevention and Management in Developing Countries. Natural Resources/Water Series 5. Bangkok, Thailand.

United Nations Environment Programme. 1988. Floodplain Management. Nairobi, Kenya.

Natural Forest Management

Fearnside, P. M. 1989. "Extractive Reserves in Brazilian Amazonia: An Opportunity to Maintain Tropical Rain Forest under Sustainable Use." Bioscience 39(6):187-393.

Food and Agriculture Organization. 1982. Environmental Impact of Forestry. Guidelines for its Assessment in Developing Countries. Conservation Guide 7. Rome, Italy.

____. 1979. Mountain Forest Roads and Harvesting. Forestry Paper 14. Rome, Italy.

Goodland, R. and others. 1985. Environmental Management in Tropical Agriculture. Boulder, Colorado: Westview Press.

Lal, R. 1986. "Conversion of Tropical Rainforest: Agroeconomic Potential and Ecological Consequences." Advances in Agronomy 39:173-264.

Lamprech, H. 1989. Silviculture in the Tropics: Tropical Forestry Ecosystems and Their Tree Species and Methods for Their Long Term Utilization. Eschborn, West Germany.

Mergen, F., and J. R. Vincent, eds. 1987. Natural Management of Tropical Moist Forests: Silvicultural and Management Prospects of Sustained Utilization. New Haven, Connecticut: Yale School of Forestry and Environmental Studies.

World Resources Institute. 1985. Tropical Forests: A Call for Action. 3 Volumes. Washington, D.C.

Plantation Development/Reforestation

Anderson, D. 1987. The Economics of Afforestation: A Case Study in Africa. Baltimore, Maryland: The John Hopkins University Press.

Boland, D. J., ed. 1989. Trees for the Tropics: Growing Australian Multipurpose Trees and Shrubs in Developing Countries. ACIAR Monograph 10. Canberra, Australia.

Cannell, M. G. R., and J. E. Jackson. 1985. Attributes of Trees as Crop Plants. Huntington, United Kingdom: Institute of Terrestrial Ecology, Natural Environment Research Council.

Cernea, M. M. 1989. User Groups as Producers in Participatory Afforestation Strategies. World Bank Discussion Paper 70. Washington, D.C.: World Bank.

Food and Agriculture Organization. 1978. Establishment Techniques for Forest Plantations. Forestry Paper 8. Rome, Italy.

Office of Technology Assessment. 1984. Technologies to Sustain Tropical Forest Resources. Washington, D.C.: United States Congress.

Postel, S., and L. Heise. 1988. Reforesting the Earth. Worldwatch Institute Paper 83. Washington, D.C.

Weber, F. R., and C. Stoney. 1986. Reforestation in Arid Lands. Arlington, Virginia: Volunteers in Technical Assistance.

Irrigation and Drainage

Ahmed, Y. J., ed. 1982. Environmental Guidelines for Irrigation in Arid and Semi-Arid Areas. Nairobi, Kenya: UNEP.

Hillel, D. 1987. Efficient Use of Water in Irrigation: Principles and Practices for Improving Irrigation in World Bank-Financed Projects. World Bank Technical Paper 80. Washington, D.C.: World Bank.

Plusquellec, H. L., and T. Wickham. 1985. Irrigation Design and Management: Experience in Thailand and Its General Applicability. World Bank Technical Paper 40. Washington, D.C.: World Bank.

Tillman, R. E. 1981. Environmental Guidelines for Irrigation. Millbrook, New York: Botanical Garden Cary Arboretum (for the United States Agency for International Development).

World Bank and the United States Agency for International Development. 1989. Irrigation Training in the Public Sector: Guidelines for Preparing Strategies and Programs. Washington, D.C.

Livestock and Rangeland Management

Child, R. D. and others. 1984. Arid and Semiarid Lands: Sustainable Use and Management in Developing Countries. USAID/NPS Natural Resources Expanded Information Base Project. Morrilton, Arkansas: Winrock International.

Dregne, H. E. 1983. Desertification of Arid Lands: Advances in Desert and Arid Land Technology and Development. Volume 3. New York: Harwood Academic Publishers.

Glantz, M. H., ed. 1977. Desertification-Environmental Degradation in and Around Arid Lands. Boulder, Colorado: Westview Press.

Little, P. D., and M. M. Horowitz. 1987. Lands at Risk in the Third World: Local-level Perspectives. Boulder, Colorado: Westview Press.

Nelson, R. 1988. "Dryland Management: The 'Desertification' Problem." Environment Department Working Paper 8. World Bank, Washington, D.C.

Pratt, D., and M. Gwynne. 1977. Rangeland Management and Ecology in East Africa. Huntington, New York: Krieger Publishing Company.

Sandford, S. 1983. Management of Pastoral Development in the Third World. Chichester, United Kingdom: John Wiley and Sons.

Sponsner, B., and H. S. Mann, eds. 1982. Desertification and Development: Dryland Ecology in Social Perspective. New York: Academic Press.

Watershed Development

Food and Agriculture Organization. 1977a. Guidelines for Watershed Management. Conservation Guide 1. Rome, Italy.

World Bank. 1989. Asia Region Review of Watershed Development Strategies and Technologies. Asia Region Technical Department, Agricultural Division. Washington, D.C.: World Bank.

____. 1990. Asia Region Strategy for Watershed Development. Asia Region Technical Department. Washington, D.C.: World Bank.

Rural Roads

Cook, C. C., H. L. Beenhakker, and R. E. Hartwig. 1985. "Institutional Considerations in Rural Roads Projects." Staff Working Paper 748. World Bank, Washington, D.C.

Goodland, R., and H. Irwin. 1975. Amazon Jungle: The Environmental Impact of the Transamazonian Highway. New York: Elsevier Scientific.

McCutcheon, R. 1988. "The District Roads Programme in Botswana." Habitat International 12(1).

Low, K. S. 1984. "Effects of Road Construction on the Environment." In Technology, Culture and Development, U. A. Aziz and others, eds. Kuala Lumpur, Malaysia: Institute of Advanced Studies.

Parizek, R. R. 1971. "Impact of Highways on the Hydrogeologic Environment." In Environmental Geomorphology, R. Coates, ed. Binghampton: State University of New York.

United Nations Environmental Programme. 1986. Environmental Guidelines for Rural Roads. Nairobi, Kenya.

ANNEX 8-1

Sample Terms of Reference (TOR)
An Environmental Assessment of Agroindustry

Note: Paragraph numbers correspond to those in the Sample Terms of Reference (TOR) Outline in Annex 1-3; additional paragraphs are not numbered

7. Task 1. Description of the Proposed Project.

(a) Characteristics of the processing plant(s): location, general layout, size, capacity and life-span.

(b) Preconstruction and construction activities of plant(s) as well as any marine terminals, deepwater ports, pipelines, or roads required.

(c) Operation and maintenance activities, including:

- Handling operations for raw materials and the form in which they are to be introduced into the facility process, as well as the off-loading, conveying, pretreatment, and storage operations. Whenever possible, information should be supplied on the source and quantities of the pollutants likely to be produced during each operation.

- Types of processing operations. For example, process control measures should be specified, as variations in the process may result in different amounts and quality of polluting substances being released to the environment.

- Waste disposal and pollution control measures categorized by continuous, batch, intermittent and emergency (spills, accidents), especially waste minimization (source reduction or recycling) schemes.

- Transportation requirements and extent to which facilities are owned, operated or supported by the proposed agroindustry. Transportation requirements for raw materials (e.g., live animals, vegetables and fruits, plant residues) should be evaluated.

(d) Supply source for raw materials and the extent to which this source is owned, operated or supported by the proposed agroindustry.

8. Task 2. Description of the Environment.

(b) Biological environment: fauna, including aquatic organisms (particularly fish); ecologically important or sensitive habitats, including parks or preserves, significant natural, cultural or historic sites, etc.; any biological factors likely to influence the supply of raw materials to the facility (e.g., pests).

10. Task 4. Determination of the Potential Impacts of the Proposed Project. Special studies may be necessary to obtain the following information:

(a) Effluent studies to define the extent of potential pollutant loading to receiving waters and to develop alternatives for providing appropriate levels of treatment. The quality and quantity of potential effluents and water pollutants -- processing waters, cooling waters, sanitary waters, leachates from solid waste disposal areas, stormwater runoff -- should be determined. Chemical characterization depends on the food products being produced, but at a minimum should include: temperature, pH, total suspended solids (TSS), oil and grease, biological oxygen demand (BOD), and chemical oxygen demand (COD). Sanitary wastewater and waters from slaughterhouses should be characterized for nitrates and fecal coliform.

(b) The quality and quantity of air emissions, including sulfur dioxide, carbon dioxide, nitrous oxides, toxic pollutants and particulate matter.

(c) The quality and quantity of solid wastes and the potential impacts from their disposal.

(d) Potential noise levels from the facility.

(e) The potential impacts from transportation should be assessed. When the plant is to be sited in a remote or sparsely populated area, the impacts from planned and unplanned in-migration into the area. This should include effects on the natural resources of the area (e.g., clearing of forests for agriculture) and socio-economic impacts (see Chapter 7).

(f) The effects of facility development on aesthetics and visual quality.

(g) Ability of the community or government to provide emergency response services for accidental release of dangerous chemicals (in most instances agroindustries pose no significant threat of this), and availability of medical facilities and trained personnel to respond to medical emergencies.

(h) Raw material handling and waste disposal specifications (to minimize the potential for disease transmission, especially in slaughterhouses and tanneries).

(i) Potential for unplanned development to result from project and the possible environmental and socio-economic effects of this.

17. Consulting Team. Members of the team might consist of people with the following specializations: environmental impact assessment; sanitary engineering for the evaluation of air and water quality, estimation of potential pollution problems from the processing plant and planning for water and air pollution control systems; acquatic ecology; plant ecology, wildlife and conservation ecology (if there is potential for negative impacts on important species or habitats); rural sociology; agronomy/livestock management, as appropriate for assessment of the impact of the production system for raw materials.

ANNEX 8-2

Sample Terms of Reference (TOR)
An Environmental Assessment of Dams and Reservoirs

Note: Paragraph numbers correspond to those in the Sample Terms of Reference Outline (TOR) in Annex 1-3; additional paragraphs are not numbered

7. Task 1. Description of the Proposed Project: general layout, size and capacity (dam and reservoir specifications, location of outlets, etc.), and life-span of the dam and reservoir.

8. Task 2. Description of the Environment. Assemble, evaluate and present baseline data on the relevant environmental characteristics of the study area, including watershed, site of the dam and reservoir and downstream areas, especially floodplain, and biological environment (particularly fish resources).

10. Task 4. Determination of the Potential Impacts of and Impacts on the Proposed Project. Potential impacts to be assessed include:

(a) Social and ecological effects of reservoir inundation (loss of agricultural, forestry and grazing land, population resettlement, effects on wildlife and wildlands, etc.).

(b) Effects on the hydrology and water quality of the river (and where relevant, the estuarine, coastal and marine resources).

(c) Effects on river fisheries and potential for creating a reservoir fisheries resource.

(d) Impacts of altering river flow regimes on the ecology of the floodplain, and the economic activities/land use on the floodplain (agriculture, livestock production, etc.).

(e) Impact of altering water supply on urban, industrial, and rural users.

(f) Potential environmental and social impacts by planned and unplanned (spontaneous) in-migration into the area.

(g) Potential for increased incidence of water-borne and water-related diseases.

(h) Impact on terrestrial and aquatic wildlife, by creation of the reservoir, disruption of migration routes, alteration of floodplain ecology, and population impacts.

(i) Effect of existing and predicted land use in the watershed on the functioning and longevity of the dam and reservoir.

17. Consulting Team. Members of the team should consist of people with the following specializations: environmental planning and management; hydrology; terrestrial ecology (plant ecology, forestry and wildlife); acquatic ecology and fisheries; watershed management; soil science and geology (where relevant); public health, particularly speciality in water borne and water-related diseases; rural sociology.

ANNEX 8-3

Sample Terms of Reference (TOR)
An Environmental Assessment of Flood Protection

Note: Paragraph numbers correspond to those in the Sample Terms of Reference (TOR) Outline in Annex 1-3; additional paragraphs are not numbered

7. Task 1. Description of the Proposed Project: scheme's general design, capacity and degree of protection for various flood levels.

8. Task 2. Description of the Environment. Assemble, evaluate and present baseline data on the environmental characteristics of the study area, including watershed areas, the sites of flood control structures and floodplain areas.

(a) Physical environment: surface and groundwater hydrology (annual peak discharge, recurrence intervals of various peak discharges, and peak stages for various discharges).

10. Task 4. Determination of the Potential Impacts of the Proposed Project. Particular attention should be paid to the following aspects:

(a) Effects of the flood control dam: direct environmental impacts of the dam construction and reservoir inundation; effects on fisheries resource (creation of a reservoir fisheries, loss of downstream fisheries); effects on water quantity and quality; effects on floodplain ecology.

(b) Effects of flood control structures and measures (e.g., levees, dikes and channelization measures) on: aquatic ecology, particularly fish resources; hydrology, including groundwater recharge, and water quality; plant and animal ecology of the floodplain.

(c) Socio-economic impacts on populations in inundation area and downstream (floodplain dwellers, urban population, etc.) through: land use changes; impacts on water-related economic activities (e.g., fisheries, transportation, etc.); health effects (e.g., increased incidence of water-borne and water-related diseases).

17. Consulting Team. Members of the team should consist of people with the following specializations: environmental planning and management; fisheries and/or aquatic ecology; hydrology, watershed management and forestry (for upstream effects); terrestrial ecology and wildlife ecology, etc. (for impacts in the inundation area and on the floodplain); rural sociology.

ANNEX 8-4

Sample Terms of Reference (TOR)
An Environmental Assessment of Fisheries

Note: Paragraph numbers correspond to those in the Sample Terms of Reference (TOR) Outline in Annex 1-3; additional paragraphs are not numbered

7. Task 1. Description of the Proposed Project: construction activities (e.g., port and harbor facilities, roads, fish processing plants, etc.); operation and maintenance involved in fishing and fish processing activities.

8. Task 2. Description of the Environment. Assemble, evaluate and present baseline data on the environmental characteristics of the study area. Include information on any changes anticipated before the project commences.

(a) Physical environment: geology, topography and soils (for aquaculture projects); climate and meteorology; hydrology, water quality; coastal and oceanic parameters.

(b) Biological environment: flora (aquaculture and freshwater capture projects); fauna (fish stock assessment -- target fish and bait fish, and for aquaculture, fingerlings); rare or endangered species (plant and animal); ecologically important or sensitive habitats, such as mangroves and presence of parks or preserves, and significant natural, cultural or historic sites, etc.; non-target species of commercial importance.

(c) Socio-cultural environment (include both present and projected where appropriate): population; land and water use; fishing rights; control over fishing rights; community structure; employment and availability of labor; role of fisheries in the local economy; public health; distribution of income, goods and services; marketing/use patterns of local fisheries; presence of producer organizations; customs, aspirations and attitudes (including traditional gender roles, e.g., role of women in fisheries); cultural properties; tribal peoples and/or other culturally or economically marginal groups; other planned development activities.

10. Task 4. Determination of the Potential Impacts of the Proposed Project. Among the issues to be investigated are the following:

(a) Potential for overfishing (e.g., of target, bait and non-target species; in capture fisheries).

(b) Type of fishing equipment currently in use and proposed for the project. Vessels, nets, catch limits proposed should be reviewed to avoid overexploitation of fish stocks, harvest of non-target species, or deterioration/destruction of habitat (e.g., seagrass beds, coral reefs, etc.; capture fisheries).

(c) **Impacts of construction and operation of fish ponds on local hydrology (surface and groundwater), water quality, vegetation and wildlife (aquaculture).**

(d) **Relationship/interaction between industrial/commercial fisheries and artisanal, small-scale fisheries; social and economic effects of increased fishing on local markets, local nutrition levels, income of local fisherfolk).**

(e) **Impacts of construction and operation of port and harbor facilities (e.g., destruction of local wildlife habitat, coastal erosion, sediment loading of local water bodies, water pollution, overtapping of fresh water supplies for processing requirements, solid waste disposal problems, etc.).**

(f) **Pollution from fish processing facilities.**

(g) **Pollution from fishing or collection ships.**

(h) **Extent of external threats to fishery resource from pollution and deterioration of fish habitat from agricultural, industrial or municipal sources.**

17. **Consulting Team. Members of the team should consist of people with the following specializations:**

Capture Fisheries: fisheries biology, rural sociology;

Aquaculture: aquaculture, terrestrial ecology (soil science, plant ecology, wildlife ecology, and for fish pond projects), rural sociologist; and

Fish Processing: wastewater/pollution management specialist, environmental impact assessment (with experience in port and harbor facilities and fish processing plants).

ANNEX 8-5

Sample Terms of Reference (TOR)
An Environmental Assessment of Natural Forest Management

Note: Paragraph numbers correspond to those in the Sample Terms of Reference (TOR) outline in Annex 1-3; additional paragraphs are not numbered

7. <u>Task 1. Description of the Proposed Project</u>: location; general design and extent of activities; construction activities (e.g., roads, sawmills, etc.); forestry operations and forest products processing.

8. <u>Task 2. Description of the Environment</u>.

(a) Physical environment: topography (specifying the watershed function of the proposed project site); climate and meteorology; surface and groundwater hydrology; estuarine, coastal and oceanic parameters (where relevant).

(c) Socio-economic environment: the presence of tribal peoples and/or other culturally or economically marginal groups; land and resource use; system of land tenure and resource use rights; employment and labor availability; extent of use of non-market forest products; existing and planned development activities, particularly water developments (dams, irrigation systems for which the project area serves as a watershed).

10. <u>Task 4. Determination of the Potential Impacts of the Proposed Project</u>. Among the areas requiring particular attention are the following:

(a) Siting: impacts on important flora and fauna (particularly threatened and endangered species); impacts on important habitats for ecosystems.

(b) Operation and management of forestry activities: short term and long term effects of logging on forest structure and biodiversity (species composition of vegetation and wildlife), including effects of the logging method on the capability of the natural forest species to regenerate naturally (sustainability of production); direct environmental effects of logging on the soil, non-target vegetation, hydrology, and local surface water quality, etc. (sustainability of environmental services); long term effects of logging through loss of nutrients from the system, hydrologic changes, disruption of wildlife habitat, etc.; effects of use of pesticides/herbicides; effects on existing uses of the forest (for agriculture, grazing, harvesting of wood and non-wood forest products); social and economic effects of logging camps and road construction teams.

(c) Transport and processing: effects of logging roads (direct effects through erosion, disruption of wildlife) as well as induced effects of increased population influx; negative effects of construction of processing plant and processing operations (particularly pollution).

17. Consulting Team. Members of the team should consist of people with the following specializations: forestry (harvesting and extraction); forest ecology; wildlife management; hydrology; watershed management; rural sociology.

ANNEX 8-6

Sample Terms of Reference (TOR)
An Environmental Assessment of Plantation Development/Reforestation

Note: Paragraph numbers correspond to those in the Sample Terms of Reference (TOR) outline in Annex 1-3; additional paragraphs are not numbered

7. Task 1. Description of the Proposed Project: construction activities (e.g., roads, sawmills, etc.); operation and maintenance activities in forestry and forest products processing.

8. Task 2. Description of the Environment.

(a) Physical environment: topography (specifying the watershed function of the proposed project site); surface and groundwater hydrology.

(c) Socio-economic environment: the presence of tribal peoples and/or other culturally or economically marginal groups; system of land tenure and resource use rights; availability of potential labor force and need to bring in outside laborers; cultural characteristics, customs, aspirations and attitudes, including existing uses of trees and experience with tree planting and management.

10. Task 4. Determination of the Potential Impacts of the Proposed Project. Among the areas requiring particular attention are the following:

(a) Siting: disruption of traditional access to and uses of the land and resources; negative effects on important flora and fauna resources.

(b) Plantation establishment and management: effects of site preparation; clearing existing vegetation, using mechanical means and burning, etc.; use of pesticides/herbicides; quantification (when possible) of positive and negative short-term and long-term environmental effects of establishing plantations/tree planting (including on soil fertility, soil erosion rates, hydrology, wildlife, etc.); positive and negative social effects of plantation/reforestation activities (increased or decreased accessibility of fuelwood, fodder and forage resources, increased employment opportunities, etc.); social and economic effects of importing laborers from outside.

(c) Transport and processing: potential effects of logging roads (direct effects through erosion, disruption of wildlife) as well as induced effects of increased population influx; problems associated with construction and operation of processing facility (especially pollution).

17. Consulting Team. Members of the team should consist of people with the following specializations: environmental assessment; forestry (plantation, harvesting and extraction); forest ecology; wildlife management; hydrology; watershed management; rural sociology.

ANNEX 8-7

Sample Terms of Reference (TOR)
An Environmental Assessment of Irrigation and Drainage

Note: Paragraph numbers correspond to those in the Sample Terms of Reference (TOR) outline in Annex 1-3; additional paragraphs are not numbered

7. Task 1. Description of the Proposed Project: general design and extent of irrigation and drainage works (specifications of dam and reservoir, size of command area, etc.); size of catchment area; operation and maintenance of irrigation works.

8. Task 2. Description of the Environment.

(c) Socio-economic environment: land use (including current crops and cropping patterns); land tenure and land titling; present water supply and water uses (including current distribution of water resources if irrigation systems already exist in area); control over allocation of resource use rights.

10. Task 4. Determination of the Potential Impacts of and Impacts on the Proposed Project. Potential impacts to be assessed include:

(a) Project Location: resettlement of people; loss of forest land; loss of agricultural land (cropping and grazing); impact on flora and fauna; impact on historic and cultural sites; effects on water resources outside and inside command area.

(b) Project Design: disruption of hydrology; drainage problems; design of dams and other structures; crossings for people and animals.

(c) Construction Works: soil erosion; construction spoils (disposal of); sanitary conditions and health risks associated with construction camp and workers coming into area; social and cultural conflicts between inported workers and local people.

(d) Project Operation: pollution by agrochemicals; impacts on soils (waterlogging, salinization, etc.); changes in groundwater levels inside and outside command area; changes in surface water quality and risks of eutrophication; incidence of water-borne and water-related diseases.

17. Consulting Team. Members of the team should consist of people with the following specializations: environmental impact assessment (with extensive experience in irrigation); rural sociology.

Depending on the baseline data needed and the mitigating measures proposed, the team may also include some of the following disciplines: agronomy; hydrology; terrestrial ecology (plant, forestry and wildlife as appropriate to the ecology of the irrigation site and adjacent areas); aquatic ecology and fisheries; soil science.

ANNEX 8-8

Sample Terms of Reference (TOR)
An Environmental Assessment of Livestock and Rangeland Management

Note: Paragraph numbers correspond to those in the Sample Terms of Reference (TOR) outline in Annex 1-3; additional paragraphs are not numbered

7. Task 1. Description of the Proposed Project: type and extent of livestock/rangeland activities; construction activities (e.g., processing facilities).

8. Task 2. Description of the Environment.

(a) Physical environment: soils (including signs of and susceptibility to erosion); surface and groundwater hydrology (particularly information on waterpoints or water availability for livestock -- location, condition and existing uses of water points and potential for developing alternative points).

(b) Biological environment: flora (natural vegetation, as well as range condition and trend, degree of degradation of vegetation around waterpoints, carrying capacity of the land for livestock, etc.); fauna (livestock: size, composition and condition of herds, seasonal distribution and movement of livestock animals; wildlife (species: numbers, habitat requirements, migratory routes, interaction with livestock, etc.); livestock health and presence of any biological factors that would potentially affect livestock numbers or health (e.g., disease vectors, noxious plants); ecologically important or sensitive habitats, including parks or preserves, significant natural, cultural or historic sites, etc.

(c) Socio-economic environment (include both present and projected where appropriate): land use, including competing demands for the rangeland and water resources; land tenure (including rights to grazing lands and water resources); control over allocation of grazing and water use rights; intensity and timing of use of waterpoints; marketing/use of livestock and livestock products; presence of producer organizations; customs, aspirations and attitudes (including traditional gender roles).

10. Task 4. Determination of the Potential Impacts of the Proposed Project. Among the issues to be investigated are the following:

(a) Effect of the project on the numbers of livestock, range condition and trend, range carrying capacity and range ecosystem.

(b) Where mechanical site improvement and burning of vegetation is called for, effect on soil, flora, fauna and hydrology.

(c) Where use of pesticides/herbicides are called for, potential for pollution, negative health impacts and negative effects on flora and fauna.

(d) Impact on the quality and quantity of water resources (surface and ground).

(e) Impact of project activities on other resource users (non-target pastoralists, agriculturalists, wildlife).

(f) Social and economic effects of the proposed project on the livestock owners or hired herders directly involved, distribution of benefits between different sectors of the society and genders, and effects on non-target populations.

17. Consulting Team. Members of the team should consist of people with the following specializations: rangeland ecology; soil science; wildlife and/or plant ecology, as appropriate; hydrology; rural sociology.

ANNEX 8-9

Sample Terms of Reference (TOR)
An Environmental Assessment of Rural Roads

Note: Paragraph numbers correspond to those in the Sample Terms of Reference (TOR) outline in Annex 1-3; additional paragraphs are not numbered

7. Task 1. Description of the Proposed Project: location of roads; type of roads and expected volume of use; construction activities.

8. Task 2. Description of the Environment.

(b) Biological environment (of road site and potential area of influence of the road); ecologically important or sensitive habitats, including parks or preserves; significant natural, cultural or historic sites, etc.

10. Task 4. Determination of the Potential Impacts of the Proposed Project. The engineering plans should reflect "best practice" in road alignment and construction to ensure that potential negative environmental impacts are minimized (e.g., through measures to prevent soil erosion risk, to ensure proper drainage, and provide for waste disposal such as of cut and fill material, used oil, etc.). The EA should verify that this is the case.

The EA should focus on the potential for negative environmental and social impacts caused by planned and unplanned (spontaneous) in-migration of people: clearing of forest lands for agriculture; increased pressure on fuelwood, fodder and water resources; social disruptions and conflicts; threat to wildlands and important wildlife species, etc.

17. Consulting Team. Members of the team should consist of people with the following specializations: rural sociology; human geography; terrestrial ecology (wildlife, plant and conservation ecology).

CHAPTER 9

POPULATION, HEALTH AND NUTRITION; URBAN DEVELOPMENT; TRANSPORTATION; WATER SUPPLY AND SEWERAGE

Population, health and nutrition loans or credits typically do not have complex or significant adverse environmental impacts and are not covered in the project guidelines. The general topic of public health and safety is covered in this chapter, but it will have equal relevance to the projects discussed in Chapters 8 and 10. TMs managing population, health and nutrition projects should keep in mind that they are not necessarily altogether free of impacts. In fact, proper management of hospital waste, which could result from a variety of health projects, can be a complicated undertaking.

Transportation is normally the Bank's second or third largest sector in terms of yearly loan and credit totals. In 1990, highways accounted for more than one-half of the total. Railway projects and ports and waterways were the other large categories. Roads and highways, inland navigation, and port and harbor facilities are the transportation categories included in this edition of the Sourcebook, as being the types of projects among those frequently financed that raise significant environmental issues. Other topics such as airports, urban mass transit, and railways, will be added if demand for EA guidelines warrants. This chapter also discusses some of the general environmental issues pertaining to urbanization which are relevant to a number of project topics.

The guidelines on wastewater collection, treatment, reuse and disposal have general applicability to any size system. However, in keeping with the Sourcebook's focus on loans or credits which may have significant environmental impacts, the emphasis is on projects in urban rather than rural areas. Water supply is not specifically covered, but the elements of water supply projects most likely to have significant impacts, dams and reservoirs, are discussed in detail in Chapter 8. The section on "Oil and Gas Pipelines" in Chapter 10 may be useful in environmental review of large water transmission mains. "Water Resource Management," in general, is discussed in Chapter 2.

PUBLIC HEALTH AND SAFETY

General Issues

1. Like environment, public health and safety is a field unto itself and covers a wide spectrum of issues throughout nearly all Bank sectors. Broadly defined in the context of the overall goal of improving the quality of people's lives, it pertains directly to nearly every project and indirectly, to all. Even when more narrowly defined, public health and safety still encompasses a vast array of traditional considerations such as air and water pollution and provision of health care; and largely because of their general relevance, widespread health benefits can be promoted by integrating complementary health or safety components into non-health specific projects.

2. However, the very breadth of public health and safety that provides an opportunity to improve generally the quality of life poses special managerial problems. For instance, the institutions responsible for public health and safety tend to be among the weakest in governments, with limited real influence and budget except in cases of epidemics, disasters or emergencies. Additionally, responsibilities may fall into the proper oversight of several government agencies and so fall between institutional cracks. Consequently, there is a genuine risk that some of the broader and long-term issues may be overlooked. This section of the Sourcebook, therefore, focuses on identifying and overcoming special problems and on suggestions for setting realistic goals.

Bank Policy, Procedures, and Guidelines

3. There is no formal Bank policy on public health and safety. General policy on the subject is mentioned primarily in operational directives, guidelines and technical papers. The overriding policy emphasizes that in design, implementation and eventual operation, projects should improve the quality of life of project beneficiaries or at the least, not diminish it. At least as much general policy direction is contained implicitly in procedures. In some cases, implicit health and safety policy is declared in public forum, as in the case of disposal of hazardous or toxic wastes: the Bank will not finance shipment or disposal of hazardous or toxic wastes from one country to another among borrowing countries. (The Guidelines noted below deal with safe handling and disposal of wastes generated within a country from its own operations.) In other cases, policy weaves its way into Bank practice through study seminars that later are formulated into policy, e.g., the seminar on Disaster Prevention and Mitigation which was issued later as an Environment Department Discussion Paper (Kreimer and Zador 1989). Still others with public health and safety implications fall entirely under a different category, such as women in development, social forestry or adjustment lending, and are not labelled specifically health or environment.

Relationship to Bank Investments

4. Since the very nature of Bank work is directed toward improving the quality of life in developing countries, from its inception the Bank has been involved with public health and safety issues. Prior to the creation of the Population Health and Nutrition Department (PHN) in 1980, the Bank dealt with such issues through components primarily in agriculture, rural development, water supply, sanitation and urban development projects, as well as through various special programs. The majority of these components

dealt with construction of health clinics, promotion of health education, and upgrading/rehabilitation of maternal/child health centers, hospitals and other health facilities. With the creation of the PHN Department, the Bank was able to expand these activities to free-standing projects.

5. After the Bank's reorganization in 1987, activities in population, health and nutrition were merged with those in education to form the Population and Human Resources Sector (PHR). The priorities in PHR lending are: population and family planning, including maternal and child health; financing health care delivery, especially at the regional and district levels; tropical disease control and AIDS; nutrition; and information, education and communication (IEC).

6. These lending, policy and research activities within the Bank are supported by several collaborative initiatives with other organizations. These include: Research and Training in Tropical Diseases Programme (focusing on pertinent socioeconomic research and seven major tropical diseases that are important in developing countries -- malaria, schistosomiasis, filariasis, African trypanosomiasis, Chagas' disease, leishmaniasis, and leprosy; Human Reproduction Programme; Safe Motherhood Initiative; Task Force for Child Survival; Global Programme on AIDS; U.N. Administrative Committee on Co-ordination/Sub-Committee on Nutrition (ACC/SCN); and Onchocerciasis (River blindness) Control Programme.

7. Collaboration may occur within the Bank as well, of course. Projects or components such as housing, water supply, irrigation, drainage, roads, solid waste removal, etc., whose justification may not be health improvement explicitly, have impacts on health and can be significantly enhanced by public health and safety components or even parallel projects where the situation merits.

8. For example, a component to provide stormwater drains (to help control mosquitos or other vectors that spread several tropical diseases) or a health education program can be added to an existing housing project to maximize the benefits from the improvements made to the physical environment. Examples of recurrent problems that can be addressed directly in this way include:

- contamination of drinking water with fecal, agricultural, industrial and other waste;
- unsafe handling, transport, storage and disposal of hazardous wastes from hospitals, industry, commerce and agriculture (see section on "Industrial Hazard Management," concerning industrial/hazardous materials safety);
- spread of diseases such as schistosomiasis (through construction of dams) or malaria (through construction of rural roads or poorly maintained dams and irrigation systems);
- contamination of food with pesticides and herbicides;
- effects of natural disasters; and
- deterioration of basic water, shelter, and health services in the public and private sector.

9. In addition, public health and safety components can have beneficial impact indirectly on pervasive problems, especially:

- population growth outstripping provision of adequate food or of basic community services;
- macroeconomic policies that inadvertently promote malnutrition, stress or the spread of diseases;
- physical and mental stress from persistent congestion, noise, lack of privacy, fear of accidents or crime, drownings, fire and landslides -- all directly associated with urban slums or squatter settlements or settlements on other marginal land;
- physical stress, particularly to women and children, from hours spent daily fetching water, firewood, and getting to and from other work; and
- serious diseases resulting from chronic low-dose exposure to household cooking and heating fumes and from illnesses associated with urban over-crowding.

10. World Bank projects, furthermore, are in key positions to make significant contribution toward eradicating the six major diseases of developing countries: malaria, schistosomiasis, hookworm, diarrhea, respiratory ailments, and malnutrition. (A seventh category -- accidents -- should be added to these six. While accidents do not account for the same degree of illness and death as the others do, they are an important consideration in nearly all projects.) The task is very difficult, and made more so for not being able to control exposure outside the geographic or temporal confines of projects. Malaria and schistosomiasis control are especially problematic because they entail long-term control of mosquito and snail populations. (By comparison, hookworm and especially guinea worm control could quite likely be achieved in the short term.) Diarrhea, respiratory ailments and malnutrition involve behavioral change as well as improvement in physical circumstances, and so are even more difficult to control. Nevertheless, as has been discussed above, even small -- but effective -- components (e.g., education about the means of disease transmission) integrated consistently and systematically into Bank projects make cumulative contributions that over time become effective.

Special Consideration

Unique Aspects of Public Health and Safety Projects

11. The following list provides a summary of considerations for public health and safety issues, as well as reminders of how their treatment may differ from projects in other sectors.

- Measurable improvement in many public health and safety issues depends on behavioral change and that takes time. This means prevention is more cost effective than cure. Preventive measures should be integrated into project design well before appraisal. This should be the primary focus of all public health and safety work. Even with the best preventive measures, health problems often manifest themselves well beyond the project cycle. Therefore, public health and safety considerations also call for follow-up after the usual period of Bank involvement in projects, as do environmental considerations.

- Many health and safety improvements become perceptible only after the last of a set of interrelated variables falls into place and should be evaluated accordingly. In public health and safety programs, it is common to read incorrectly the lack of early success as evidence of failure.

- Behavioral change is labor-intensive for Bank staff as well as field staff, imposing on projects practical administrative constraints -- particularly to time and budget.

- Some governments are reluctant to admit or publicize existent or potential public health and safety problems for political or economic reasons.

- Staff in ministries or agencies administering projects with public health and safety components often are not trained in environment or health.

- The agency or public utility responsible for executing projects with potential health impacts may not have in-house capability (i.e., trained staff or infrastructure), or accountability for environment or health.

- Strengthening individual institutions requires interministerial and interagency cooperation, a managerial practice that is not efficient in many developing countries.

- Agencies may balk at taking on administrative activities outside their normal scope of work, especially if it extends beyond the project cycle's funding.

- Funding as well as executing agencies may be reluctant to take on administratively cumbersome (e.g., labor-intensive) projects or components, even if justified on social grounds.

- Funds for public health and safety, and related social aspects, often come from local budgets rather than from loan funds and thus may be vulnerable to competing demands for limited resources.

- Beneficiaries themselves may not understand public health and safety problems and therefore prefer investments in more immediate needs, such as street lights, schools and public markets.

Women and Public Health and Safety

12. No matter what combination of services make up a typical project, women figure predominantly as key participants and beneficiaries. Whether healthy or sick, women support families, manage households, fetch water and fuelwood, and care for children. Their specific needs will therefore require consideration as a high risk group and as a primary audience. Sustained health improvements from project interventions, in a large portion of bank projects, hinge on education and behavioral change, factors that frequently depend on women. The role of women in ensuring the success of a project or component should not be underestimated, and their opinions and participation should be actively solicited.

Guidance for Environmental Assessments

13. There are two principles that underpin realistic planning for public health and safety programs. First, public health and safety improvements should be designed in the context of the multiple causes of disease and death so that those in separate projects (or even in other sectors) will be integrated. That is,

reduction in disease generally depends on changes in both physical environment and behavior, with all the attending complexities, and individual interventions should be designed within that context. Second, interventions may not have a perceivable impact over the short-term (particularly on some of the most prevalent diseases), but should be done anyway and in such a way as to lay the needed foundation for future success. An example of these two principles is a hygiene education component in a clinic project that can strengthen and be strengthened by a drainage renewal component of a housing project.

Health Components as Part of Project

14. A positive health impact can be included in a traditional infrastructure project in at least the following three ways:

(a) ***By improving facilities (even without a specific health project or component) where behavioral change is not necessarily required and therefore, follow-up is not required.*** Examples are:

- bacteriological/chemical contamination controlled by change in source or treatment of water;
- filariasis eliminated by improved waste disposal and drainage;
- respiratory diseases reduced by providing more efficient cooking/heating facilities and ventilation;
- Guinea worm eliminated by changing or protecting the source of water; and
- public safety enhanced by redesign of public buildings and housing in areas prone to earthquakes, hurricanes or floods.

(b) ***By including public health and safety components which require only moderate programs and follow-up.*** Local conditions can often be substantially improved by relatively simple interventions. Moreover, improvements tend to take place in the short term and to continue, assuming proper maintenance of the facilities. Examples are:

- intestinal worms substantially reduced through provision of sanitation facilities combined with training in hygiene;

- accidents (such as drowning and burns) eliminated by modifications to the living environment and training in safety.

(c) ***By including public health and safety components which require extensive programs and follow-up.*** In many cases, a positive impact can actually occur within months but requires long-term follow-up to be sustained and to allow time for behavioral change. Examples are:

- diarrhea substantially reduced through the provision of water and sanitation facilities combined with training in personal and public hygiene;

- malnutrition substantially reduced through extensive long term health education provision of nutritional supplements, and improvement to living conditions and water supply.

Free-Standing Public Health and Safety Projects

15. In some circumstances, a positive impact is best achieved by a project directly addressing public health and safety. Consequently, an environmental assessment should alert project officers to situations where a sole health component may not be effective, and a free-standing project is required. For example, controlling diseases such as schistosomiasis, malaria and those resulting from industrial pollution require substantial long-term efforts and are better tackled on a larger scale. The environmental assessment should note disease prevalence and recommend separate interventions, e.g., as a free-standing health project or a government program executed by another agency. Whether to design a health component to a non-health project or to design a free-standing project with a health impact is largely determined by the feasibility of including long-term administrative supervision in the project.

Steps to Prepare Environmental Assessment

16. **Gather baseline health data.** Gather the information necessary to determine the current health status of the population living in the project area; specifically, to identify existing and predicted problems and to define a suitable response. This may begin simply with a list of the top ten causes of illness and death of the target population. If there are other projects planned or in progress (Bank or non-Bank) within the region, they should be identified.

17. Gathering and interpreting information could be performed by the entity responsible for managing the anchor project or the health/safety component. However, borrowers such as municipalities or public utilities may have no expertise in the health or education sector and no resources to provide assistance on long-term health components. The work will then most likely be performed by consultants engaged by the agency or through cooperative arrangements between the agency and local non-government organizations (NGOs). Where project organizations lack expertise in the health field to undertake extensive health-related activities, the EA may need to specify an alternative supervisory agency or begin the process of institution strengthening.

18. **Relate basic data to sectors.** A clearer definition of needs is produced by relating the top ten causes of illness and death to specific deficiencies, e.g., in housing, water and sanitation. Again, relate this analysis to other health and safety projects or activities in the target population.

19. **Determine source of existing and anticipated problems and type of project intervention or alternative.** Carry the analyses a step further by comparing the main sources of the problems with the focal points of prevention, such as water quality control, waste disposal, and education.

20. **Select the intervention required to achieve the fullest health impact from the investment.** From a short list of possible health/safety projects or components identified, final selection can be made on the basis of their feasibility of implementation and potential impact. Wherever feasible, integrate components with ongoing health and safety activities. Each proposed initiative should be accompanied by terms of reference.

21. In view of administrative practicalities, health and safety projects or components can be defined as having short- or long-term impacts. A short-term impact is one that can be demonstrated within the project cycle, but may require follow-up. Demonstration or pilot projects, if coupled with long-term follow-up outside the project cycle, qualify as long-term impact components.

22. The critical factor in each case is the capability of the project entity to administer, operate and maintain the component. Usually, free-standing projects provide for their own infrastructure. A primary necessity of the health/safety component is obtaining administrative support, perhaps through collaboration with an agency outside the project. In general, the more complicated the administration and the higher the requirement for follow-up, the greater is the need to consider public health and safety interventions as free-standing programs.

23. **Determine complexity of interventions chosen.** Analyze the possible interventions to determine whether they are appropriate within the project. Considerations to address include time span, labor and material requirements, follow-up, etc.

24. **List possible components.** With an understanding of problems, potential interventions and their complexity, it should be possible to derive a list of projects or components. Again, the more complicated the intervention or component, the more likely it should be viewed as a separate project. In such cases, where project interventions would be administratively cumbersome, a component may consist of a demonstration project, or securing collaboration with another agency may be called for.

25. **Preparing proposal for component.** The list of possible projects or components can then be worked into a specific proposal and Terms of Reference prepared.

26. As routine parts of EA, all projects should be assessed for the possibility that they may impair public health and increase the risks of accidents. To help reduce these risks, all contracts might carry appropriate clauses, such as: "The contractor shall screen all candidates prior to hiring for their health status...shall appoint a permanent accident prevention officer...shall provide health facilities free to local people...." This ensures that the project will provide appropriate public health and safety measures and will engage experienced officials to administer them.

Conclusion

27. This section has focussed on public health and safety as a general issue. These principles should be augmented by reading the other relevant sections of the Sourcebook: the chapter on social and cultural issues and the project-specific discussions on agriculture, industry, and so forth.

28. Again, it must be emphasized that in public health and safety programs, success cannot be defined or measured by standard parameters, e.g., rates of return or construction schedules. Ultimate benefits from public health and safety measures are difficult or impossible to quantify because correlation with standard measures is complicated by many interrelated variables interacting simultaneously -- and also because the unit measured is human life.

29. When designing projects, remember that nearly every Bank project can strengthen and be strengthened by a public health and safety measures within a project or as a separate component; Bank projects are well-situated to make a pivotal contribution to health and safety generally in developing countries; and even small components -- if they are well-designed and integrated systematically into projects -- can cumulatively contribute significantly to human wellbeing.

ENVIRONMENTAL CONSIDERATIONS FOR DEVELOPMENT PROJECTS IN URBAN AREAS

1. Cities play a key role in the development process. They are, in general, productive places that make more than a proportionate contribution to national economic growth. However, the very process of urban growth often brings with it a deterioration in surrounding environmental conditions. As a locus for population growth, commercial and industrial activity, cities concentrate energy and resource use and waste generation to the point that both man-made and natural systems are overloaded and the capacities to manage these systems are overwhelmed. This situation is exacerbated with rapid urban population growth. The resulting environmental damages or costs threaten the continued productivity of cities and the health and quality of life of its citizens. Cities have become major "environmental hot spots" that urgently require special attention in regional and project EAs and in environmental planning and management at the metropolitan regional scale.

2. Urban systems and services (e.g., water supply, sanitation, public transport and roads) are increasingly congested due to population, commercial and industrial growth coupled with poor urban management. Natural resources (water, air, forests, minerals, land) vital to the cities' economic development and to future generations are lost or misused through inappropriate urban policies. The radius of impact of cities on resources lying far beyond their boundaries is steadily increasing. Furthermore, urban areas are inundated in their own wastes and choked on their own emissions as a result of inadequate pollution control and waste management policies and practices.

3. Many negative impacts are associated with the conditions described above. The greatest health risks in many developing country cities are still linked to the traditional problems of excreta disposal. At the same time, there is increasing concern about the health risks of modernization due to toxic wastes and emissions, traumas (traffic and other accidents, violent deaths), and urban stress. The spatial scale of these impacts stretch from the household level through the community, the urban area, and in some cases the region beyond. The impacts of greatest concern are still often found at the household and community scale, and are related to deficiencies in urban infrastructure and services. Urban dwellers, and particularly the urban poor, bear the brunt of deteriorated environmental conditions through health and productivity losses and diminished quality of life. Resource development costs are rising (e.g., the cost of new water supplies) as economically accessible high-quality resources are depleted. Emissions related to regional and global environmental problems are increasingly generated in urban areas or as a result of urban demand (for example, urbanization alone could be a major factor in world energy demand over the next generation).

Environmental Impacts of Urban Growth

Pollution from Urban Wastes

4. Undoubtedly, the major pollutant of concern in developing country cities is human excreta. Fewer than 60 percent of the urban population in less developed countries has access to adequate sanitation, and only one third are connected to sewer systems. Where sewerage does exist, rarely do cities provide any treatment. Similarly, the collection and disposal of household garbage is a persistent problem of local government. Only half the urban residents benefit from collection services, and the

municipal solid wastes that are collected end up in open dumps or drainage systems. Industrial wastes are a particular concern in cities without adequate waste management capacity, as it is difficult to monitor discharges and ensure that hazardous wastes do not end up in city sewers or landfills. As a consequence, the sophisticated facilities and sites needed to treat and dispose of hazardous wastes are not yet in place and used in most developing countries.

Urban and Household Air Pollution

5. Air pollution is a growing problem in large cities that have poor natural ventilation and significant mobile or stationary emissions. In many cities, conditions are worsening from year to year as emissions from fuel use and industry increase. For example, vehicle fleets and associated emissions are expected to grow 5-10 percent per year in developing countries with most of the growth concentrated in major cities. Urban emissions represent a significant and growing share of greenhouse gases and those responsible for the destruction of the ozone layer. As serious as ambient air pollution is in many large cities, a distinction should be made regarding indoor air pollution, which throughout the Third World is possibly a more serious and ubiquitous problem. In the housing environment, a major concern is the indoor combustion of high polluting traditional fuels for cooking and heating, which often results in daily exposure to high levels of toxic compounds. Occupational exposure to indoor air pollution and other toxics is also a frequent problem, especially in small manufacturing enterprises. From a public health viewpoint, the key problem is the total exposure of an individual over time from all urban sources -- ambient, household, and occupational.

Water Resource Problems

6. Urbanization, coupled with inseparable industrial development, has profound impacts on the hydrological cycle -- both quantitative and qualitative. Available water resources around cities are being depleted and/or degraded to the point that marginal costs of supplies are increasing substantially. These cost increases arise both from the need to develop new, more remote sources and from increased treatment requirements resulting from deteriorating water quality. Depletion results largely from inadequate pricing policies and conservation measures. Overpumpage of groundwater in many cases leads to land subsidence with attendant damage to urban structures, to a lowered groundwater table, and in many cases to salinization problems. Improper disposal of urban and industrial wastes contributes to water quality deterioration of both surface sources and aquifers. In the case of groundwater, contamination is a particularly serious problem that can lead to the permanent loss of valuable high-quality drinking water supplies. The land surface impermeabilization in urban areas dramatically changes the runoff hydrograph, leading to higher peaking and more frequent flooding, and direct ground water recharge is often reduced. At the same time, urban runoff is a major source of non-point pollution. Water pollution problems in lakes, coastal and marine waters can result in amenity losses (recreational opportunities and tourism revenues), fisheries depletion, and health problems associated with contact recreation and fish and shellfish contamination.

Energy Production and Consumption

7. Cities inevitably involve the consumption and, to a varying extent, the production of energy for home, business and transportation use. The very scale of urban energy use may lead to significant pollution problems, already discussed above. Excessive use is common, often the result of inefficient

energy policies, such in the countries of Eastern Europe where energy consumption per unit GDP is four times that of their industrialized neighbors. The projected increases in energy use in cities of the less developed countries, based on population and economic growth forecasts and present inefficient consumption patterns and high-polluting fuel choices, portend worsening environmental impacts at local, regional and global scale and risks of major accidents. Appropriate fuel and energy pricing, improved energy planning and management practices, and energy saving technologies are needed as fundamental pieces of energy conservation strategies. Cities also affect natural heat balances. The heat generated by energy usage, including motor cars, combined with daytime heat trapping by urban structures and the slower release of stored heat at night, creates a heat island effect that can raise in-city temperatures by 5° to 10° C. In hot regions and seasons it is almost invariably unfavorable, adding to discomfort and even contributing to deaths (e.g., among the elderly) when it aggravates a general heat wave. Thermal inversions are commonplace above urban areas, trapping industrial emissions and products of fuel combustion and contributing to the formation of noxious photochemical "smog".

Degradation of Land and Ecosystems

8. As cities grow, the failure of urban land markets induces inappropriate land development and pressure is exerted on surrounding natural ecosystems. Impacts may include the loss of wetlands and wildlands (together with their rich genetic diversity and hydrological buffering capacity), coastal zones, recreational areas, and forest resources (particularly due to accelerated deforestation to meet urban demand for firewood and charcoal). Urban development can negatively impact on downstream watersheds through increased runoff and erosion. Watershed degradation can also occur far from the city, as for example when major water supply or hydropower projects are built hundreds of kilometers away, or when firewood and charcoal are brought from such distances.

Occupation of Hazardous Areas

9. The uncontrolled periurban development observed in many developing country cities leads to the occupation of low-lying lands, floodplains and steep hillsides by squatter and low-income groups. In addition to the land degradation that often results, the people populating such areas are themselves exposed to increased health hazards from flooding, landslides and mudslides, and erosion, and their dwellings and the surrounding community infrastructure are vulnerable to accidents, damage and collapse. Health risks may also result from living in close proximity to high-polluting industries or from industries working with hazardous wastes (e.g., Bhopal).

Loss of Cultural Property

10. Often overlooked when considering environmental impacts is the degradation of cultural property in cities, whether it be historical or living monuments. Air and water pollution are primary culprits which accelerate the decay and destruction of these cultural resources. The impacts are felt in the loss of cultural heritage and of tourism revenue.

Features of Urban Environment Requiring Special Consideration in EA

Dynamics of the Urbanization Process

11. Although the pace of urban growth is not without historical precedent, the sheer magnitude of population growth is overwhelming. Cities such as Mexico City and Sao Paulo are adding a half to three-quarter million new inhabitants yearly. The now-evident slowdown in the rate of growth of the megacities will have little impact on the numbers of new dwellers seeking jobs, housing and services. On the other hand, the rate of growth of secondary and intermediate cities shows no sign of abating. Many developing countries will be faced with an explosion of urban population in cities of size classes from 20,000 up to 500,000 or one million, and should now be deciding on how to provide the needed urban infrastructure. Past government policies to influence or control rural-urban migration and industrial location have been misdirected and, by and large, have failed. Furthermore, the growth of many large cities is now dominated by natural increase rather than migration. Future efforts to rationalize urban growth should focus on family planning, and on neutral locational policies that seek to ensure equal opportunity of access for individuals and enterprises to efficient urban infrastructure and services in all cities and towns, rather than subsidized services for the privileged few in primary cities.

Scale of Waste Concentration and Resource Consumption

12. Pollution prevention and control efforts are hampered by the scale of urban growth and the rapid concentration of urban wastes and emissions of all types. Scale here is defined as the product (per unit area) of population density times per capita waste generation (or resource use). The problem is exacerbated by economic development as the quantity of urban wastes generated per capita tends to also increase rapidly with increased per capita income. The scale of the problem exceeds the capacity of local government to collect, treat and dispose of municipal sewage and solid wastes, the capacity of authorities to control dangerous wastes and emissions, and the capacity of nature to assimilate all such wastes. The challenge is to find ways of reducing urban waste generation in addition to building efficient, effective, and affordable waste management and pollution control programs.

13. An analogous scale problem exists on the input side as a result of the concentrated resource consumption taking place in urban areas. Urban demand for energy, water, food, minerals, timber and fuelwood, and other resources often has impacts on distant watersheds and forests. Increasingly important factors in the sustainable development and use of these resources are urban demand management, technological change and resource substitution.

Cross-Media and Cross-Sectoral Issues

14. Most attempts to deal with urban environmental problems are partial, fragmented and compartmentalized. Four classes of fragmentation are common.

(a) Cities tend to give attention to problems perceived by the community (e.g., waste collection) while ignoring the externalities they occasion (e.g., siting and impacts of waste disposal).

(b) Programs that focus on a single environmental medium fail to protect other media (e.g., sewage treatment plants clean up the wastewater stream but produce large quantities of sludge that have to be dealt with; while disposing of sewage sludge and solid wastes on land may prevent surface water pollution problems at the expense of groundwater and soil quality). Tracing such cross-media effects is often complex.

(c) The failure to coordinate sectoral programs contributes both to the above cross-media problems and to the loss of resources spent on ineffective actions (e.g., investments in surface drainage without parallel improvements in solid waste collection and disposal; attempts to develop fisheries in waters polluted by municipal wastes).

(d) Jurisdictional conflicts impede effective actions or give rise to crucial gaps in prevention and control efforts.

Deteriorated Living Environment and Public Services

15. In general the quality of the built environment and the provision of infrastructure services are adversely affected by the pace of urban growth and the inadequacy of institutional responses. Both investment and operational capacity are inadequate to meet the challenge. The results are manifest in substandard housing and ventilation, lack of water and sanitation, trash piles in streets and drains, proliferation of disease-carrying insects and rodents, traffic congestion, lack of open and green spaces, noise and stress, and vulnerability to natural disasters, among others.

Impacts on the Poor

16. As the physical environment in and around cities deteriorates, those most affected are the urban poor, who make up 30 to 60 percent of the urban population in developing countries and whose numbers are growing. Foremost among the environmental concerns are the health problems of these inhabitants, whose substandard living environment does not protect them from human and other wastes and indoor air pollution. Intraurban studies confirm that the mortality and morbidity from gastroenteric and respiratory infections and malnutrition are significantly higher for the urban poor than for other urban residents. So too are the resulting costs of health care and the productivity losses. For the poor, the main environmental priorities remain improved housing and the provision of basic water and sanitation services at affordable costs. Moreover, in the rapidly industrializing cities of the developing world, there is evidence that the poor are also at significantly greater risk from chronic and degenerative diseases associated with modernization, as well as traumas and psycho-social problems stemming from environmental stress. Among the urban poor, there are several particularly vulnerable groups -- children, women, adolescents, cottage industry workers, and the elderly. Structurally, these groups are particularly exposed because they do not have the political strength to push for environmental improvements, the economic ability to invest in mitigating measures and pay for services, and the knowledge about alternatives.

Spatial and Intergenerational Impacts

17. The analysis of spatial scale of environmental impacts is crucial to the understanding of urban environmental problems. These problems are also related to the provision, or lack thereof, of key urban infrastructure and services as is also indicated in the figure. The consideration of spatial scale of impacts pinpoints several important issues for developing countries.

(a) Health impacts are greater and more immediate at the household or community level and tend to diminish in intensity as the spatial scale increases.

(b) Equity issues arise in relation to the provision of basic services at the household or community scale (see discussion above on intraurban health differentials and poverty) and intertemporal externalities at the regional and global scale -- particularly the intergenerational impacts implicit in non-sustainable resource use and global environmental issues.

(c) Levels of responsibility and decision making should correspond to the scale of impact, but existing jurisdictional arrangements often violate this principle.

Some Causes of Urban Environmental Degradation

Insufficient Knowledge and Understanding

18. The complexity engendered by the unique features of the urban environment described above makes it difficult to adequately understand and analyze the relationships between problems, impacts, underlying causes and curative actions. The capacity to plan and implement responses to the urban environmental problems cited above, and the ability to build a political consensus, is hampered by insufficient knowledge of both the problems and the processes of environmental degradation. Environmental data for urban areas are usually lacking, and there is no broadly accepted analytical framework for understanding the problems, ranking them and designing environmental protection programs. Robust conceptual models in this area urgently need to be agreed upon and disseminated, whether they be simple, pragmatic planning models, expert systems, or more sophisticated environmental quality models for prediction. An analytical framework to guide data collection and problem analysis is provided (see Table 9.1 at the end of this section).

Economic Policies

19. **Inadequate pricing of both resources and services.** Inefficient resource pricing (e.g., land, water, energy, food) undercuts attempts at resource management, and results in the inability to conserve scarce resources and the excessive use of natural resources leading to eventual ecosystem collapse. While this problem is not unique to urban resource management, the scale of urban consumption and the strong political pressure to keep prices down for urban consumers argues for particular attention to urban demand management. Inappropriate pricing of urban services results in inefficient allocation and inequitable distribution, as manifested by:

- urban poor without services despite a certain willingness and ability to pay
- subsidized services for the rich
- misallocation of funds between capital and recurrent expenditures
- inability to recover costs and finance services locally leading to greater dependence on fiscal transfers from central government

20. Consideration may be given to using subsidies (e.g., soft credits for service connection, life-line rates, or user cross-subsidies) as transitional measures to shield the urban poor from potential negative short-term impacts of structural adjustment, but in such cases they should be open, transparent, and well-targeted.

21. **Other sources of economic distortion.** Hidden subsidies, poorly devised or enforced taxes, interest rates, exchange rates, and regulations can all lead to damaging distortions in the use of environmental resources, both physical and natural. Apart from direct pricing as mentioned above, other types of distorting subsidies are common. For example, protective tariffs for manufactured products that are energy inefficient or highly polluting, failure to charge transport or stumpage fees for urban-bound fuelwood or charcoal production leading to ineffective forest management, and the adoption of an interest rate far below the cost of funds for municipal loans or housing finance that undermine the very sustainability of the financing mechanism. Excessive regulatory control of land markets results in unaffordable land prices and leads to the occupation and degradation of ecologically sensitive lands as well as the location of the poor on hazard-prone lands.

Weaknesses in Environmental Planning and Management

22. **Deficient legal and regulatory framework.** Few countries have as yet established the comprehensive environmental protection legislation and regulations that are needed to manage urban environmental problems. The law should provide a clear assignment of jurisdiction and responsibility for monitoring and enforcement, as well as specify funding mechanisms for enforcement and control activities. It should also be consistent with current scientific knowledge and establish clear objectives, policies and processes for promulgating regulations and standards, and for setting and maintaining sanctions commensurate in value with environmental damage incurred. Regulations and standards should be kept simple, transparent and easy to apply. In many cases, significant environmental improvement could be achieved within existing legislation and regulations, if it were not for an almost total absence of monitoring and enforcement capacity.

23. **Jurisdictional complexity** in urban areas results from the multiplicity of actors (national and regional sector agencies, state or provincial governments, local government) with responsiblities for key functions unassigned, overlapping, underlapping or poorly defined. Often there is failure to match level of authority with extent of responsibilitiy or scale of impact. There is also a need to separate normative and operational functions. Thus local government may have an important role in regulating pollution from industry (particularly small-scale manufacturing), while at the same time municipalities are major polluters who should be regulated by regional or national authorities.

24. **Weak organization and management.** Weak urban management hampers efforts to improve environmental conditions in cities. This cuts across both planning and operational responsibilities -- for municipal waste management and pollution control, land management, drainage, traffic control and municipal finance. Environmental concerns are rarely incorporated into the planning process. In many countries, separate sector agencies provide individual environmental services with inadequate coordination between agencies. Cost recovery for services is seldom adequate even for operation and maintenance, leaving no funds for financing new investments for unmet demand. Inefficient management of municipal and industrial waste treatment and disposal operations makes pollution control a major priority in most urban regions. Public institutions lack the resources to do the job, but there is little effort to involve the

private sector in the provision of urban environmental services under conditions that would favor efficient operations.

25. **Ineffective land use controls and property rights systems.** Property rights issues contribute to the degradation of land and natural resources and to the inequitable distribution of such resources. Few countries have resolved basic land tenure problems that impede mobilization of household and community resources for basic environmental improvements. This is especially the case for periurban squatters and central-city slumdwellers. Water rights are another problem area, as urban industrial use of groundwater is often in direct conflict with the efficient development of regional water resources. The environmentally sound development of urban land use is often thwarted by the failure of local land markets due to over-regulation in addition to the lack of land information systems and the absence of policies and actions to guide urban expansion away from areas poorly suited to urban development.

Lack of Environmental Awareness and Political Will

26. Sometimes the link between environmental cause and effect is so remote in time or space that it is not recognized, or if it is, social and private interests may diverge. Political leaders often focus on immediate and highly visible problems, leading to short-term "bandaid" solutions. Often, unfortunately, it takes an environmental disaster (e.g., Bhopal or a series of floods) to stimulate profound change. Public information campaigns and consultative processes are needed that identify the key environmental priorities, build the political will to deal with them, and help build community capacity to participate in the solutions. Too often environmental education is totally lacking in the formal educational system, is not seen as a function of environmental authorities, and is done a disservice by the press which tends to focus on sensationalistic treatment of environmental topics rather than real local priorities.

Opportunities to Improve Urban Environmental Management

Policy Interventions

27. **Economic policies** can be a powerful tool for redressing the environmental imbalances occurring in and around cities. Pricing resources and services at their true marginal social cost would improve resource conservation and utilization. In comparison with other environmental policies, pricing measures tend to have a low cost and in many cases are fiscally positive. Subsidies for environmentally destructive activities should be removed. On the other hand, economic incentives are also needed to reduce excessive reliance on regulation and investment programs. Such administrative measures as pollution charges, tax incentives, and targeted subsidies may be useful.

28. **Urban hygiene policies** are also needed that focus on such areas as shelter improvements, water supply and sanitation, urban and industrial waste management, and accidents prevention. Other important areas are drainage, traffic management, energy management, and preventative health care programs. Policies should be aimed at increased affordability and coverage coupled with cost recovery.

Environmental Management and Planning

29. A balanced **legal and regulatory framework** is necessary to support investment programs and economic policy instruments. While deregulation is sought in some areas (e.g., land markets), stricter

regulation and enforcement is required in other (e.g., discharge standards). Environmental legislation may need to be updated to take into account new scientific knowledge, phase in environmental regulations and standards in accordance with investment and enforcement capacity, and create new consultative mechanisms between levels of government, business, and the public.

30. **Land use management** is crucial for mitigating the impact of urbanization on land and other resources. Land registration and information systems, property tax systems, land tenure arrangements, and spatial planning are all important management tools that are not used widely enough in developing country cities. Serviced urban land is a scarce good, and its use and expansion should be guided to spatially appropriate areas. The failure of urban land markets leads to such environmental problems as land degradation, loss of fragile ecosystems, and occupation of hazardous areas. Urban capacity to plan for and manage land use is crucial for environmental protection in such cases, as is attention to both formal and traditional property rights systems. Land development regulations and standards that constrain the provision of serviced and affordable urban land and housing should be removed or modified.

31. **Natural resource management.** There are significant interactions of urban areas with the surrounding natural resource base both in terms of excessive urban demand for resources or through resource degradation resulting from urban actions or inactions. Manifestations of environmental impacts are common in water resources, forests and coastal zones. These occurrences point to the need for integrated regional planning and management of key natural resources around major urban areas. Furthermore a legal and regulatory framework for natural resource development is needed that can feasibly be implemented, clearly defines the relationship between legal and administrative authorities, clarifies property rights, and is comprehensible to the public. While local government may not have a major role to play in natural resource management, the influence of cities on resource development cannot be overlooked.

32. **Improved urban waste management** is a particular municipal function that demands urgent attention. The development of institutional capacity and the introduction of sound management practices are needed to achieve expanded coverage for low-income groups, efficient service provision, effective disposal operations. The potential role of the private sector in providing waste management services is greatly undervalued and should be explored further.

33. **Environmental planning** means integrating environmental concerns into the overall urban planning process. This requires the incorporation of environmental information, policies, standards, techniques and monitoring into coordinated strategic action planning at the city level. Spatial, cross-media, intersectoral and intertemporal factors must all be accounted for in environmental planning.

Institutional Improvements

34. A necessary improvement is in the area of institutional capacity for enforcement actions. Monitoring and enforcement are crucial whether adopting a command and control strategy, economic incentives, or a mix of the two for environmental protection. Since environmental problems cut across political boundaries, jurisdictional arrangements and responsibilities need to be spelled out. Typically, national environmental authorities are responsible for establishing environmental policies and regulations and providing technical and financial assistance to local governments for program implementation. Regulatory and enforcement actions to control private polluters are taken by state and local governments

or special regional authorities created for river basin management or waste disposal districts. Municipal government is almost universally assigned responsibility for domestic waste collection and disposal, and in failing to manage these activities appropriately is itself a major polluter. A clear distinction must be drawn between such polluting activities of local government and those corresponding to local government control of industrial discharges (e.g., control of emissions in urban areas, of discharges into municipal sewers, and of hazardous waste disposal in municipal landfills). Regional, state or national authorities must be able to enforce regulations on municipal waste disposal by cities. Finally, the role of community groups, environmental NGOs, and the private and informal sectors in urban environmental management must be recognized and encouraged through appropriate enablement policies.

Investments and Financing

35. Increased capital investment is also necessary for addressing the growing backlog of urban environmental infrastructure needs. There should be greater emphasis on the use of appropriate and cost-effective technologies for environmental protection, together with reduced dependence on the public budget for such investments through the application of beneficiary charges. Public/private partnerships are an especially promising approach. The application of the "polluter pays" principle can reduce the cost to government of curative programs. At the very least, self-financing of environmental services through user and polluter charges must be sufficient to pay for recurrent costs of operations. Ideally, these charges should be linked to the long-run marginal costs of building, maintaining and operating environmental infrastructure.

Environmental Education and Public Participation

36. To apply political pressure for environmental action, affected urban groups must understand the causes of their problems and be able to vocalize their concerns. Typically, though, there is little public awareness about the sources and effects of urban ecological decay. Further, target populations are rarely involved in the planning and implementation of environmental infrastructure and services. This is particularly true for vulnerable groups such as women, children and the elderly. Consciousness-raising is needed through bringing information about environmental hazards and alternatives to those most affected by environmental externalities. Awareness can motivate affected groups to participate in the process of environmental management. Environmental education is required to inform key urban groups about the causes of ecosystem and resource deterioration, as well as potential solutions at the household, community, city and regional levels. This can be accomplished through formal mechanisms (e.g., public schools, media campaigns) and informal, targeted approaches (e.g., training community leaders, neighborhood extension agents). Public involvement can be broadened and deepened by encouraging a variety of actors (community movements, business organizations, consumer protection groups, political parties, youth groups and others) and by opening the planning and implementation processes to public participation.

Table 9.1. Analytical Framework for Urban Environmental Problems

Manifestations	Impacts	Causes	Cures
Deteriorated living environment and services	Health impacts - infectious and parasitic diseases - malnutrition - accidents - stress, drugs, violence - acute and chronic toxicity - genetic effects - cancer Lost urban productivity Degradation of the natural environment - resource loss - amenity loss	Substandard housing Lack of water and sanitation Disease-carrying insects and rodents Indoor air pollution Excreta laden water/soils Trash dumping Noise/stress Traffic congestion Natural disasters	Appropriate housing and land development regulations Housing finance Provision of affordable infra-structure and plots Provision of affordable water sanitation, solid waste services at cost Targeted subsidies Improved efficiency and effectiveness of infra-structure and service provision Pollution control Community participation
Ambient air pollution	Localized primarily in large cities - health problems - lost aesthetic, cultural, recreational values - property damage (including historical monuments At present more of metropolitan area problem than regional/ global problem	Urbanization/industrialization Fuel/energy pricing and urban energy demand Vehicle ownership Space heating Use of highly polluting fuels: - leaded gases - high-sulphur lignite	Pricing of industrial and energy inputs Regulations and standards Emission charges Monitoring and enforcement Energy conservation Technological interventions - scrubbers, baghouse filters - vehicle emissions control - fuel substitution
Indoor air pollution	Chronic obstructive lung disease	Biofuel use for domestic cooking and heating	Pricing of commercial fuel and energy

Table 9.1. Analytical Framework for Urban Environmental Problems (continued)

Manifestations	Impacts	Causes	Cures
Indoor air pollution (continued)			
	Acute respiratory infections Low birth weight and associated problems Cancer	Passive smoking Cottage industry exposure	Targeted fuel subsidies Improved housing and ventilation Public awareness
Solid waste pollution	Health hazards Amenity impacts Blocked drainage and flooding Water pollution (leachates) Air pollution (heavy metals/toxic organic compounds from incinerators and uncontrolled burning)	Inefficient management (collection and disposal) Impacts not recognized or external to community	Improved collection - expanded coverage (e.g., to low-income area via community based approaches) - efficient operations (e.g., create contestable markets to encourage private sector entry) - financial strengthening (budgeting, accounting, cost recovery) Disposal technology and management Resource recovery/recycling
Fecal contamination	Diarrheal diseases Parasitosis Malnutrition High infant mortality	Lack of basic sanitation services Excreta laden water/soils Proliferation of garbage/insects	Provision of affordable sanitation options at cost Community-based approaches Hygiene education
Hazardous waste pollution	Acutely affects groundwater, but often multimedia impacts Damage at low concentrations - Health damages (e.g., acute and chronic toxicity)	Insufficient regulations/management Pricing of inputs into into industries producing waste	Regulations, standards and changes Monitoring and enforcement capacity Licensing Waste minization - process modification

Table 9.1. Analytical Framework for Urban Environmental Problems (continued)

Manifestations	Impacts	Causes	Cures
Hazardous waste pollution (continued)			
	- Foodchain accumulation "Timebombs" (hidden dumps that build up over time)	Impacts removed in space and time	- resource recovery/recycling Treatment and disposal technology and management
Fresh water resource depletion (surface and ground water)	Sources running out Increasing marginal costs Land subsidence	Pricing policies Cultural (perception of free good) Overpumping of groundwater Irrigation policies and practices (inefficiency) Leakages from water	Improved pricing Integrated watershed management Improved technologies (e.g., wastewater reuse) Regulation of groundwater extraction Improved O & M
Fresh water resource quality degradation (surface and ground water)	Poor quality surface and ground water Health impacts Increasing marginal cost of: - potable supply - industrial supply - individual treatment Waterlogging and salinization Saline intrusion	Municipal and industrial waste disposal practices - Sewerage (water pricing, poor O & M) - Industrial wastes (input pricing, poor regulations enforcement) Urban runoff Irrigation policies and practices Overpumping of groundwater Impacts not recognized or external to the community	Pricing Policies Regulations, standards, charges Monitoring and enforcement Solid and hazardous waste management Treatment technology and operations Integrated watershed management Regulation of groundwater extraction Public education

Table 9.1. Analytical Framework for Urban Environmental Problems (continued)

Manifestations	Impacts	Causes	Cures
Lake, coastal and marine water pollution (including fisheries depletion)	Mainly occurs locally: - closed beaches and lost toursim revenues - lost aesthetics - health consequences - eutrophication - fish and shellfish contamination Reduced employment opportunities	Municipal and industrial waste disposal practices (see above) Agricultural runoff (plus related policies) Detergents Shipping/oil Litter/plastics	Water pollution regulations (municipal and industrial) Solid waste management Improved technologies (e.g., outfall design) Shipping facilities and regulations Special areas designation (e.g., marine sanctuary) Coastal zone management
Degradation of land and ecosystems	Loss of wetlands/wildlands (rich (genetic diversity, migratory birds, hydrological aspects) Coastal zone degradation Loss of recreational areas (e.g., beach fronts) Deforestation (urban firewood/ charcoal demand) Increased erosion	Shadow value of land much higher than prices Absence of land taxation and enforcement Uncontrolled urban growth - absence of planning control - lack of legal alternative to squatter developments Water pollution Solid waste disposal practices Occupation of steep zones	Appropriate incentives (prices and taxes) Affordable planning regulations; enforcement Pollution control regulation Special areas designation (e.g., nature preserves, parks, seashores) NGO-type activity in support of environmental initiative
Occupation of high-risk land (generally squatter and low-income groups) - low-lying land - flood plains - steep zones	Flooding Landslides, mudslides Erosion Health risks, accidents Property damage and building collapse	Land tenure systems Failed land markets Lack of developed lands Land regulations/enforcement	Appropriate incentives (prices, taxes, tenure, access to housing finance) Less regulation of land markets Provision of affordable infrastructure Targeted subsidies Community participation programs

Table 9.1. Analytical Framework for Urban Environmental Problems (continued)

Manifestations	Impacts	Causes	Cures
Degradation of cultural property - historical monuments - living monuments	Loss of cultural heritage Loss of tourism revenue	Local value less than international value Air pollution Solid waste management practices Lack of enforcement	Dead monuments: - pollution control - preservation/enforcement Living monuments: - historical districts - tax incentives - public education

ROADS AND HIGHWAYS

1. This section treats the impacts of roads and highways, which comprise expressways, primary roads, or byways and rural roads (see "Rural Roads" section for details on the latter). Non-paved roads can have substantial impact, often more than paving and existing roads. Loans or credits to build, improve or rehabilitate roads and highways are almost exclusively done as transportation sector loans or specific road and highway investment projects. Relatively few loans or credits have been approved recently for the construction of highways along totally new alignments, where the full range of environmental impacts may be expected. Rehabilitation and maintenance projects are numerous, however, reflecting the widespread problem of inadequate road system maintenance (World Bank 1988). Sectoral transportation improvement or rehabilitation loans or credits addressing this problem normally include upgrading of some number of unpaved secondary roads to paved standards. This poses a two-phased risk to the environment -- first constructing, to non-paved standards, then returning later for paving on existing alignment -- and should be avoided. Reconstruction of bridges and roadbeds to reduce flooding or flood damage can be a major component of road and highway improvement projects.

Potential Environmental Impacts

2. Socioeconomic benefits provided by road and highway projects include all-weather reliability, reduced transportation costs, increased access to markets for local produce and products, access to new employment centers, employment of local workers on the project itself, better access to health care and other social services, and strengthening of local economies. However, new roads and highways can produce complex negative impacts. The impacts of improvement, rehabilitation and maintenance projects, although usually more limited, can still be significant, not only on natural resources and systems but also on the social and cultural environment.

3. Direct impacts of road and highway projects result from construction, maintenance, and traffic use. The most significant construction-related impacts are those related to clearing, grading or roadbed construction: loss of vegetative cover; foreclosure of other land uses; modification of natural drainage patterns; changes in groundwater elevation; landslides, erosion, stream and lake sedimentation; degradation of vistas or destruction of cultural sites; and interference with movements of wildlife, livestock, and local residents. Many of these impacts can arise not only at the construction site but also at quarries, borrow pits and materials storage areas serving the project. In addition, adverse environmental and sociocultural impacts can occur in both construction and maintenance projects as a result of air and soil pollution from asphalt plants, dust, noise from construction equipment and blasting; use of pesticides; fuel and oil spills; trash and garbage; and, on large projects, the presence of a non-resident labor force.

4. Direct road and highway use impacts may include: increased demand for motor fuels; accidents with and/or displacement of non-motorized methods of transport; increased air pollution, noise, roadside litter; injury or death to animals and people attempting to cross roadways; health risk and environmental damage from accidents involving hazardous materials in transit; and water pollution from spills or accumulated contaminants on road surfaces.

5. A wide variety of indirect negative impacts have been attributed to road and highway construction or improvement (see Table 9.2 at the end of this section for examples). Many of them are primarily sociocultural and are discussed in more detail in the sections on "New Land Settlement" and "Induced Development" in Chapter 3. These include visual degradation by roadside billboards; impacts of unplanned land development induced by the project; disruption of local land ownership by speculation; construction of new secondary roads; greater human access to wildlands and other natural areas; and labor force migration and displacement of subsistence economies.

Special Issues

Induced Development

6. Many of the direct impacts on natural systems, historical and cultural resources, and right-of-way land uses can be avoided by judicious route selection. It is much more difficult to manage the impacts of new development and penetration into natural areas that may be induced by road construction or improvement. Usually that task falls to other agencies, which may not have been involved in planning the project and which, if they are at the local government level, may be ill-prepared to cope with induced development. New industry tends to locate where land is available and infrastructure exists; highway corridors are logical choices. Roadside commercial development takes place in response to speculation that improved access and greater visibility will bring more customers. Urban sprawl is a phenomenon of larger scale. In the absence of an overall plan, it consists of the expansion of urban areas outward along transportation corridors through industrial, commercial and residential developments.

7. Because of its unplanned nature, induced development proceeds without comprehensive consideration of impacts. Other infrastructure, especially that needed for waste management, may not exist. Social services may become overloaded. The land tenure of low-income landholders and indigenous peoples may be jeopardized by abrupt increases in local land values. Natural resources formerly protected from unplanned exploitation simply by their inaccessibility may become accessible and thus unprotected. Natural systems, visual amenities and historic and cultural resources may be disturbed. These individual developments also generate traffic, possibly overloading the very roads and highways which led to their existence in the first place.

Loss of Agricultural Land

8. Prime agricultural land, relatively level and well-drained, provides an ideal alignment for roads, and many are located on it. The loss of land to the right-of-way itself may be relatively insignificant and is usually taken into account in deciding whether or not to proceed with a project. However, the phenomenon of induced development, coupled with increasing land values along roads, can lead to conversion of large tracts of agricultural land which were not considered in planning. Such conversions may turn out to have negative impacts on national programs for sustainable agriculture and food self-sufficiency as well as on the viability of the local agricultural economy.

Deterioration or Loss of Ecologically Sensitive Areas

9. Road construction through wildlands, parks, natural forests and other undeveloped rural areas will inevitably lead to their conversion to other land uses, unless there is popular local support for conservation or preservation combined with effective management and law enforcement. This combination has proven elusive in most developing nations.

Interference with Customary Local Transportation Methods

10. Pedestrians, animal drawn vehicles, and pedalled vehicles are important types of traffic on roadways in many countries, especially local roads and roads leading to major market towns. Upgrading of unpaved rural roads to paved standards that does not take into account the volume of such traffic will lead to unacceptable levels of accidents and displacement of slower modes of transport. An adequate number of safe crossings and separate or parallel restricted right-of-way for slow traffic should be incorporated into road and highway projects if there is existing or latent demand for non-motorized modes of transportation in the area.

National/Global Implications

11. Road and highway construction may increase demand for motor vehicles and fuels and lubricants. If these must be imported, balance of payments problems may be exacerbated. Local and regional air quality may deteriorate, and contributions of greenhouse gases will increase. The EA should address these implications in project alternatives (see below).

Project Alternatives

12. There are no alternatives to roads that fulfill the functions of providing relatively fast, cheap land transportation. Air, rail, and water transport are more likely to complement than to substitute for roads or highways. Alternatives to construction of a new roads or expansion of existing ones which deserve consideration in transportation planning from an environmental point of view include: improvements in traffic management and public transportation on existing roads, rail systems for freight or passengers, and increased investment in non-motorized transportation, at least, for short distances.

13. Alternatives which should be investigated in planning and designing an individual project include alignments to avoid valuable or sensitive resources and alignments that do not provide access to wildlands or other areas which should remain in their natural state.

Management and Training

14. Experience shows that road maintenance is a weak area in the transportation sector. Inadequate maintenance or improper practices (with respect to chemical use, waste disposal, etc.) may lead directly to environmental impact, if vegetation, aesthetics, quality of runoff water, functioning of drains, or accident frequency are affected. Indirectly, deterioration because of inadequate maintenance will

eventually necessitate rehabilitation work, which has potentially greater impacts. Management systems and training are necessary in the maintenance area.

15. Staff in road agencies may need training in good practice to control negative impacts of road construction, maintenance and operation. They may also need training in monitoring and inspecting road projects for environmental impacts and for implementation of mitigating measures. Equipment, vehicles, and operational budgets may be lacking, and standards may be weak or vague, e.g., for air, noise and water pollution. It may be necessary to develop pragmatically defined standards, and to create a special monitoring unit.

Monitoring of Construction

16. Only construction impacts are treated here. The monitoring of immediate and longer term induced impacts and their mitigation will fall within the jurisdiction of regional development authorities, ministries of agriculture and others, according to the nature and scope of development that is induced by, or planned in parallel with, road and highway construction. If planned and integral to the project, the corresponding sections will indicate the kind of monitoring needed, e.g., land settlement or forest exploitation. If not integral to a road or transportation sector operation, certain impacts may be sufficiently important to justify a companion planning and development exercise, with its own monitoring functions.

17. Monitoring of construction impacts and of the timely and correct implementation of needed mitigating measures should be carried out according to a site-specific plan, by technical offices having mandate, personnel and the necessary capacities. In general, the most critical project elements to be monitored are the implementation and effectiveness of erosion and sedimentation control measures, disposal of debris and wastes, management and reclamation of borrow pits, and materials handling and storage areas.

Table 9.2. Roads and Highways

Potential Negative Impacts	Mitigating Measures
Direct	
1. Increased sediment in streams affected by erosion at construction sites and fresh road cuts, fills and waste dumps.	1. Protect susceptible surfaces with mulch or fabric, and plant erodible surfaces as soon as possible.
2. Soil and water contamination by oil, grease, fuel and paint in equipment yards and asphalt plants.	2. • Collect and recycle lubricants. • Avoid accidental spills through good practice.
3. Air pollution from asphalt plants.	3. Install and operate air pollution control equipment.
4. Local dust and noise.	4. • Periodically water down or lightly oil temporary roads. • Install and maintain mufflers on equipment.
5. Air and noise pollution from vehicle operation, in populated areas traversed by the highway, notably metropolitan areas or densely settled rural areas.	5. • Include physical barriers to noise in plans. • Require adherence to engine maintenance schedules and standards (or use alternative fuels) to reduce air pollution. • Enhance public transportation and traffice management capability.
6. Landscape disfiguration by embankments and deep cuts, fills and quarries.	6. • Use an architectural design to "blend" with the landscape. • Replant disfigured surfaces.

Table 9.2. Roads and Highways (continued)

Potential Negative Impacts	Mitigating Measures
Direct (continued)	
7. Landslides, slumps, slips and other mass movements in road cuts.	7. • Provide drainage works as needed to reduce risk, according to prior surveys. • Align route to avoid inherently unstable areas. • Stabilize road cuts with structures (concrete walls, dry wall masonry, gabions, etc.).
8. Erosion of lands below the road bed receiving concentrated outflow from covered or open drains.	8. • Increase number of drain outlets. • Place drain outlets so as to avoid cascade effect. • Line receiving surface with stones, concrete.
9. Roadside litter.	9. • Provide for disposal facilities. • Encourage anti-littering laws and regulations.
10. Hazardous driving conditions where construction interferes with pre-existing roads.	10. Provide in design for proper markers on roads, including lights.
11. Alteration of overland drainage and subsoil drainage (where road cuts intercept perched water tables, springs, etc.).	11. Installation of adequate drainage works.
12. Destruction of vegetation and wildlife in the right-of-way occupied by the highway.	12. Realignment where possible to detour exceptional areas, identified by prior surveys.

Table 9.2. Roads and Highways (continued)

Potential Negative Impacts	Mitigating Measures
Direct (continued)	
13. Destruction or damage of terrestrial wildlife habitats, biological resources or ecosystems that should be preserved.	13. Plan national transportation route alignment according to location of fragile, unique, etc., areas.
14. Alteration of hydrological regimes of wetlands by causeways, with harmful effects on these ecosystems.	14. • Realignment to avoid wetlands. • Installation of culverts, bridges, etc., as needed and according to criteria from prior hydrobiological surveys. • See "Wetlands" section.
15. • Interruption of migratory routes for wildlife and livestock. • Increased collisions with animals.	15. • Realign to avoid imnportant migratory routes. • Provide undergrade crossings.
16. Poor sanitation and solid waste disposal in construction camps and work sites.	16. Provide adequately located and maintained latrines.
17. Possible transmission of communicable diseases from workers to local populations and vice versa.	17. Periodic health examinations of workers with treatment when needed.
18. Creation of temporary breeding habitats for mosquito vectors of disease, e.g., sunny, stagnant pools of water.	18. Assess vector ecology in work areas and take steps where possible to avoid creating habitats.
19. Creation of a transmission corridor for diseases, pests, weeds and other undesirable organisms.	19. Set up plant and animal sanitation service and related checkpoints.

Table 9.2. Roads and Highways (continued)

Potential Negative Impacts	Mitigating Measures
Direct (continued)	
20. Poaching by construction workers.	20. Prohibit poaching under terms of employment.
21. • Dislocation and compulsory resettlement of people living on the right-of-way.	21. • See "Involuntary Resettlement" section.
• Near cities and in rich farming regions, many people can be affected.	• Locally unprecedented mechanisms and procedures may be required to arrive at equitable and adequate compensation, and a companion effort to develop the capacity may be required.
22. Obstruction of routes from homes to farms, etc., increasing travel time.	22. Provide appropriately designed and located crossings.
23. Impairment of non-motored transportation in the highway corridor due to reduced or impeded rights-of-way.	23. Include slow traffic lanes and/or paved shoulders and safe crossings.
24. Accident risks associated with vehicular traffic and transport, that may result in spills of toxic materials injuries or loss of life.	24. • Design and implement safety measures and an emergency plan to contain damages from accidental spills. • Designate special routes for hazardous materials transport. • See "Industrial Hazard Management" and "Public Health and Safety" sections.
Indirect	
25. Induced development: roadside commercial, industrial, residential, and "urban sprawl."	25. • Involve land use planning agencies at all levels in project design and EA, and plan for controlled development.

Table 9.2. Roads and Highways (continued)

Potential Negative Impacts	Mitigating Measures
Indirect (continued)	
26. Increased motorized transportation (with possible increased dependency on imported fuels).	26. Include project components to encourage use of non-motorized transportation.
27. Impairment of non-motorized transportation economy due to changes in land use and/or increased availability of motorized alternatives.	27. Include project components to stimulate local production and use of non-motorized modes of transportation.
28. Unplanned or illegal timber cutting.	28. See "Rural Roads" section.
29. Unplanned or illegal land clearing.	29. See "Rural Roads" section.
30. Illegal invasion by squatters and poachers of homelands of indigenous peoples.	30. See "Indigenous Peoples" and "Induced Development" sections in Chapter 3.

INLAND NAVIGATION

1. Projects undertaken to enhance inland navigation include dredging for channel provision, improvement and maintenance, port and harbor development (e.g., berthing facilities, barge flotilla areas, turning basins) and the construction of locks, canals, and waterways. In recent years, nearly all Bank lending in this subsector has been for rehabilitation or expansion of existing facilities.

Potential Environmental Impacts

2. Dredging is the principal practice used worldwide to improve the navigability of inland waterways. Since dredging is a well established method and remains the primary means for the development and maintenance of routes for inland navigation, there exists an extensive information base on the impacts of inland waterway projects.

3. The alteration of natural waters and the development of man-made channels can cause physical, chemical, and biological changes in the water body, leading to direct and indirect adverse impacts on related ecosystems and communities in the environs.

4. Water, land and air pollution from dredging operations, materials disposal, construction activities and increased maritime traffic can result in the release of natural and anthropogenic contaminants to the environment. (A comprehensive checklist of environmental considerations is provided in Annex 10-2, "Port and Harbor Facilities.") Since numerous dredging and materials disposal methods exist for improving inland navigation, the combinations of physical, chemical and biological effects will vary. Potential concerns include oil spills and discharges, contaminant release, habitat destruction, circulation alterations, and transportation safety. Terrestrial impacts may include contamination due to dredge materials disposal, erosion and sedimentation from hydrologic changes, and loss of habitat due to shoreline and commerce-related development. (See list of mitigating measures in Table 9.3 at the end of this section.)

5. Dredging is not the only activity having an impact on the environment; the construction of dikes and river training works involves quarrying and transportation of large quantitiites of materials.

Natural Resource Issues

Water

6. Aquatic systems can be affected by the following dredging operations: turbidity associated with the resuspension and settlement of sediments; partitioning of toxic contaminants and reintroduction to the water column; contaminant uptake by and accumulation in fish; short-term depletions of dissolved oxygen levels; modified bathymetry causing changes in circulation, species diversity, and water chemistry; and loss or modification of habitat and fisheries resources. Disposal of dredged materials may cause impacts similar to, but potentially more severe than, those associated with the dredging operation. Improved navigability will also result in increased development and water traffic that may result in accidental spills and in the discharge of oily bilge, ballast water, anti-fouling materials, and sewage. As a result of these indirect impacts, some water-based recreational activities may be precluded.

Land

7. The shorezone along the waterway can be affected by hydrologic alterations. These impacts include destruction of floodplain, increased erosion, loss of vegetation, flooding, and drainage of wetlands and mangroves. Further discussion of wetlands is provided in Chapters 2 and 3. Land disposal of dredge materials in confined and unconfined systems can impact underlying groundwater, contaminate surface runoff, and affect future land-use options. Additional terrestrial impacts would most likely result from increased vehicle traffic and industrialization related to the development of the shoreline and inland areas serviced by the port facilities.

Social and Cultural Issues

8. Inland navigation projects may upset the local cultural, ethnic, historical and religious traditions. In some cases, project acceptance and success may be hampered by local concern for the potential destruction of historical places, parks, reserves and valuable coastal zone recreational and fisheries resources.

9. Project planning and EA should provide for community involvement. Review of the impacts of creating demands on the local technical and labor resources should be conducted to prevent unacceptable pressures on limited resources. Care must be taken to limit and/or prevent adverse impacts on scarce local commodities. For example, the destruction of habitat important to a valuable local fishery could result in undesirable economic and cultural impacts. (See Chapter 7 on community involvement.)

Special Issues

Environmental Law

10. Internationally, greater attention is being given to the importance of maintaining and protecting the structural and functional integrity of coastal zone resources; therefore, any inland navigation project that may affect these resources must comply with any development restrictions. Open-ocean disposal of wastes, including contaminated dredge materials, has received considerable scrutiny in recent years. Applicable local and international regulations, such as the Oslo Convention of 1974, the Paris Convention of 1978, and the London Dumping Convention of 1972, should be followed. (International treaties and agreements on the environment and international waterways are discussed in Chapter 2).

Waterway Location

11. Many developing countries are characterized by dense human populations, inadequate potable water and sanitary waste disposal systems, intensive land use, and increasing levels of environmental degradation. Improving or developing a waterway for increasing maritime commerce and related port/harbor industries in an area currently experiencing multiple demands on environmental resources may be ill-advised, unless adequate mitigative measures are planned to ensure the proper handling of wastes from development-related activities. The decision to improve/develop a waterway is usually based on economic, geographic, physical and political parameters rather than on those of an ecologic nature. In

choosing a location, the assimilative capacity of the prevailing natural and sociocultural systems should be considered, along with accessibility, employment needs and local commerce.

Project Alternatives

12. Often, several planning, design and implementation alternatives exist for an inland navigation project. As a project progresses, Bank staff and consultants will describe the specific alternatives considered during project appraisal. Discussed below are various alternatives and considerations that can provide a framework for examination of a specific project by the environmental assessors and reviewers.

Site Selection

13. The selection of a location for enhancement of inland navigation depends on many physical characteristics of the site as well as on socioeconomic concerns. Good locations typically satisfy the following criteria.

- Physical characteristics including wind, currents, weather and siltation do not require excessive maintenance.
- Suitable land area is adequate for the processing and waste management needs of any developing industries.
- Scheduling considerations such as the spawning and migration periods of indigenous biota are not violated.
- Need for resettlement is minimal.
- Project does not compete with other highly valued land uses.
- Project activities do not adversely affect the value of an existing resource, such as a fishery.
- Construction, operation and maintenance of the waterway does not damage sensitive habitats (e.g., estuaries, mangroves); rare, threatened, or endangered species; lands and waters used by indigenous peoples; cultural properties.

Dredged Material Disposal

14. The initial screening for evaluating disposal options is a physical and chemical analysis for geotechnical character and the presence of contaminants in the sediments. Depending on the physical and chemical character of the dredged material, disposal may be confined, unconfined, or treated prior to release in open water, along the shoreline, or on land. Disposal must be in accordance with applicable regulations. Also, long-term monitoring of the dredging process and disposal may be required.

Dredging Process

15. The primary categories of dredges include mechanical, hydraulic and new innovative technologies. When selecting the appropriate dredging technology or combination of technologies, the project engineer should consider the following site-specific factors: environmental constraints associated with the physical and chemical character of the sediments; susceptibility of indigenous aquatic species to dredging-related activities; cost and availability of equipment; disposal site location and limitations; physical conditions affecting the dredge, transit and disposal locations; and interference with other users at the dredge, transit, and disposal locations.

Management and Training

16. Where applicable, a comprehensive dredging and dredged-materials management plan should be considered for inland navigation projects to ensure that development and maintenance can be carried out on schedule with minimum environmental effects. The plan should be based on: a characterization of the materials to be dredged; a vertical and horizontal profile of contaminant distribution in the channel sediments -- to define homogeneity and locate prominent "hot spots"; an evaluation of sediment behavior using different dredging equipment and disposal options; and an analysis of potential long-term effects of maintenance on human and environmental health.

17. Support for efficient pollution control and waste reduction strategies may be important for dredging and construction activities, equipment operation and materials disposal. Project engineers should be familiar with state-of-the-art equipment and disposal technologies to ensure environmentally sound management of navigation projects.

18. All project staff should receive training under "standards of practice" for occupational health and safety and emergency response. The training should include procedures to be followed in the event of accidents, spills, explosions, or fires.

19. Training for government officials charged with supervision of an environmental management and monitoring plan may be required. To assess training needs, the capacity of local institutions to assume responsibility for environmental review should be evaluated, as well as the record of legal and regulatory agencies to monitor and enforce standards.

Monitoring

20. A site-specific environmental monitoring plan that enables Bank and local government officials to manage a project and ensure compliance with environmental standards should be prepared for each individual project. Generic parameters that may require monitoring during project planning, start-up and operation are as follows: geotechnical and chemical characterization of sediments; water quality of project area and proposed disposal area; long-term chemical/physical testing of project area and disposal location; sediments and water quality; long-term monitoring of biota for the possible accumulation of contaminants; maintenance of programs to keep a high level of employee environmental awareness; and monitoring of effects of project on populations and systems in the environs of the dredged area.

Table 9.3. Inland Navigation

Potential Negative Impacts	Mitigating Measures
Direct: Dredging Process and Disposal	
1. Project location (e.g., enhancement or development of a waterway) may affect sensitive habitats and/or valuable fisheries resources.	1. Perform screening analysis of site environs and ecology, and select an area that is not adjacent to sensitive habitat and would not displace valuable fisheries resources or otherwise significantly degrade environmental quality.
2. Removal and disturbance of flora and fauna at the dredging site.	2. • Plan for minimizing impacts on local flora and fauna. • Screen for the presence of rare, threatened, or endangered species which are indigenous to the project location, and modify design to avoid or protect.
3. Interference of stationary dredging equipment with other maritime traffic.	3. Prepare a program in advance to coordinate and reduce interference to other waterway users.
4. Possible disturbance or damage to stationary installations such as underwater cables, pipelines, and outfalls.	4. • Identify and document locations of stationary installations. • Modify dredging process/disposal plans to accommodate presence of such structures.
5. Objectionable noise to nearby residents, especially at night.	5. Reduce noise level by decreasing operating level during quiescent periods in the local community.
6. Increased short-term turbidity at dredging site.	6. Reduce turbidity by efficient use of less intrusive dredging equipment, silt curtains, dredging during low flow periods.
7. Alterations of bottom surface which may be unfavorable to the success of indigenous benthic flora and fauna.	7. Plan for minimizing impact on important or sensitive benthic fauna and flora through ecological investigation during project planning.

Table 9.3. Inland Navigation (continued)

Potential Negative Impacts	Mitigating Measures
Direct: Dredging Process and Disposal (continued)	
8. Partitioning of natural and/or anthropogenic contaminants from sediments to the water column.	8. • Perform physical and chemical analysis of sediments prior to disturbance. • Locate potential "hot spots" and prepare plan to minimize sediment resuspension in these areas.
9. Modified bathymetry causing changes in tidal currents, river circulation, species diversity, and salinity.	9. Perform project area investigation, sampling, and modelling characterization and design project accordingly to minimize impacts.
10. Generation of turbidity plumes.	10. Use technologies such as temporary dams and/or barriers to lessen the transport of suspended material away from the project area.
11. Loss of shoreline integrity.	11. Evaluate shoreline geology and hydrology prior to project design to ensure deepening will not cause modifications such as slumping and increased erosion.
12. Upland disposal of dredged material could modify terrestrial habitat.	12. • Evaluate disposal options and select one with least impingement on important habitat. • Require reclamation plans for terrestrial sites.
13. Short-term air quality degradation resulting from dredging-related operations.	13. Monitor local air quality and reduce operations if unacceptable quality arises.
14. Projects may result in stress on local cultures.	14. • Evaluate local sociocultural environment prior to project implementation. • Develop specific mitigation measures with community involvement.

Table 9.3. Inland Navigation (continued)

Potential Negative Impacts	Mitigating Measures
Direct: Dredging Process and Disposal (continued)	
15. Covering of potential archaeological sites with dredge spoil.	15. Evaluate disposal area for presence of important artifacts, and modify design to salvage or protect artifacts.
16. Spills associated with increased maritime commerce.	16. Develop spill prevention and clean-up plans. Train a team to handle spills.
Indirect: Dredging Processes/Disposal	
17. Uptake and accumulation of resuspended and partitioned sediment contaminants by biota.	17. Physical and chemical analyses permits proper planning prior to project implementation, thus minimizing sediment resuspension through proper selection of dredging equipment and implement long-term biota tissue monitoring program.
18. Occupational health effects on workers from sediment handling operations.	18. • Train employees to be aware of potential occupational hazards and establish a program on safety and health which includes all of the following: • site characterization and analysis • site control • training • medical surveillance • engineering controls, work practices and personal protective equipment • monitoring and informational programs • handling raw and process materials • decontamination procedures • emergency response • illumination • sanitation at permanent and temporary facilities

Table 9.3. Inland Navigation (continued)

Potential Negative Impacts	Mitigating Measures
Indirect: Dredging Processes/Disposal (continued)	
19. Impacts of possible land disposal on groundwater, surface runoff and/or land use.	19. • If land disposal is the selected option, contaminated wastes should be contained in a structure that minimizes leachate generation and release to local aquifers. • Uncontaminated sediments have some practical uses; however, wetlands reclamation is no longer considered an environmentally sound option.
20. Transit patterns disrupted, noise and congestion created, and pedestrian hazards aggravated by heavy trucks transporting materials to/from the construction sites.	20. • Proper site selection can mitigate many of these problems, but special transportation sector studies should be prepared during project feasibility to select best routes to reduce new shorezone facilities/industries. • Develop emergency contingency plans to minimize risk of accidents during transport. • Initiate discussions on transport regulations, where none exist.

PORT AND HARBOR FACILITIES

1. Maritime transport (shipping) conveys over 82 percent of world trade; therefore, port and harbor development projects (e.g., terminals, berthing facilities, turning basins) are usually associated with long-term economic benefit for developing nations. Technological advancement in marine transport and the integration of transport by land, sea, and air have increased the complexity of port and harbor development. The dynamic character and importance of maritime transport can lead to projects such as development of approach channels, canals, waterways, and turning basins; construction of quays, breakwaters, jetties, and groins; and possibly the building of deepwater ports, prefabricated ports, offshore and relocatable terminals. Nearly all of the Bank's lending in recent years has been for rehabilitation or expansion of existing facilities.

Potential Environmental Impacts

2. The success of maritime commerce, fishing industries, and naval defense are dependent on port and harbor development; therefore, proper design, construction, and maintenance of these coastal and marine resources are important. Maritime development usually generates local environmental problems; however, development associated with sensitive estuaries or inland on freshwater rivers may yield regional-scale problems. The impacts of maritime development will differ by location because of variations in features such as geography, hydrology, geology, ecology, industrialization, urbanization and type of shipping.

3. The alteration of natural waters and the construction of man-made structures can lead to direct impacts on the water body being developed as well as direct and indirect impacts on related ecosystems and communities in the project environs.

4. Dredging operations, materials disposal, shorezone development, increased maritime traffic and vehicle traffic in the port can result in the release of natural and anthropogenic contaminants to the environment. Since numerous dredging, materials disposal, and construction methods exist for developing port and harbor facilities, the combinations of physical, chemical, and biologic effects on the medium of interest will vary. Potential aquatic impacts include oil spills and discharges; contaminant release from sediment resuspension, surface runoff, and point source discharges; habitat destruction; changes in water chemistry and circulation; occupational and public health concerns; and transportation safety. Terrestrial impacts may include contamination due to dredged materials disposal; erosion and sedimentation due to hydrologic changes caused by channel deepening and widening and shorezone development (construction of breakwaters, etc.); loss of sensitive habitat (e.g., wetlands, mangroves) due to shoreline and port-related development; and loss of existing and future land use. Air impacts can include degradation of air quality due to emissions from industrial stacks and vehicular traffic, and the generation of fugitive dusts. (A summary of all potential environmental impacts is provided in Table 9.4 at the end of this section.)

Natural Resource Issues

Water

5. Dredging and dredge spoil disposal activities for port development and maintenance can induce short- and long-term impacts on aquatic systems as follows: degradation of marine resources such as beaches, estuaries, coral reefs, and fisheries; resuspension and settlement of sediments; partitioning of toxic contaminants and reintroduction to the water column; contaminant uptake by and accumulation in fish and shellfish; increased turbidity causing decrease in light penetration and associated photosynthetic activity; short-term depletions of dissolved oxygen levels; modified bathymetry causing changes in circulation, possible saltwater intrusion to ground-water and inland surface water; altered species diversity and structure of benthic communities, and fluctuations in water chemistry; changes in shoreline structure; and loss of habitat and fisheries resources.

Similar impacts can also result from increased maritime traffic and facility development along the shorezone.

6. Improved navigability and development of port facilities will increase maritime traffic and with it, the risk of spills and the discharge of oily bilge, ballast, anti-fouling materials and sewage. The increase in shorezone development will also contribute stress to the receiving aquatic system through point discharges of sewage, process and cooling waters, and accidental releases.

Land

7. The waterfront region in the port and harbor area will be altered to suit new industry. These new industries can lead to relocation of villages, increased vehicular traffic, dust and airborne emissions from traffic and raw materials stockpiles, and contamination of surface runoff. Numerous port and harbor facilities are proximal to sensitive saltwater marshes, mangroves, and estuaries which can act as sinks for the collection of contaminated stormwater and sediment from the waterfront area. Further discussion of wetlands is provided in Chapter 2.

8. Land disposal of dredge materials in confined and unconfined systems can also affect underlying groundwater, contaminate surface runoff, and alter future land-use options. Increasing concern about the loss of wetlands and the effects of this loss on the hydrologic and biologic structure and function of biomes, will most likely preclude the use of wetlands as reclamation sites for dredged materials.

Social and Cultural Issues

9. For the most part, new or expanded port and harbor facilities in a developing nation are welcomed since these projects provide new jobs and bring an influx of commerce to the region. However, improvements, expansion, and industrialization may upset the local cultural, ethnic, historical and religious traditions. In some cases, project acceptance and success may be hampered by local concern for the potential destruction of historical places, parks, reserves and valuable coastal zone recreational and fisheries resources. Community involvement in project planning is essential.

10. During the planning and implementation phases, projections should be made of the possible impacts of increased demands on the local technical and labor resources to prevent excess pressure on limited resources. Care must be taken to limit and/or prevent impacts on scarce local commodities. For example, the destruction of a valuable local fishery or recreational beach for the development of port and harbor facilities could result in undesirable economic and cultural impacts.

Special Issues

Hazardous Materials/Cargoes

11. The shipping and handling of hazardous materials such as pesticides, explosives or pressurized gases at port facilities in developing nations could pose an unacceptable risk to human health and the environment. To protect workers and the surrounding communities, authorities should ensure that effective measures are enforced for monitoring the transport and handling of hazardous materials at the port (see "Industrial Hazard Management" section).

Maintenance Dredging

12. Maintenance dredging is performed in approach channels and harbor basins to maintain depth and width and ensure safe access for large vessels. The dredged materials from maintenance dredging typically present a greater disposal problem than deeper sediments removed during construction dredging, since surficial sediments are composed of recently deposited materials that are usually contaminated. These younger sediments usually contain natural and anthropogenic contaminants and can arrive from atmospheric fallout, erosion of local land surface and channel banks, fallout from biological activity in the water column, sediment transport from inland waters, point source dischargers, and surface runoff from the surrounding area. To mitigate potential contaminant release from the port area, the following should be addressed: proper design of stormwater handling and treatment facilities; sewage and wastewater outfalls; local land use (e.g., proximity of agricultural fields or mining operations); procedures for handling hazardous materials; and types of industries permitted to operate in the port area.

Environmental Law

13. Internationally, greater attention is being given to the importance of maintaining and protecting the structural and functional integrity of marine and coastal zone resources; therefore, any port and harbor development which may affect these resources must comply with local and/or regional restrictions.

14. Open-ocean disposal of wastes, including contaminated dredge materials, has received considerable scrutiny in recent years. Applicable local and international regulations, such as the Oslo Convention of 1974, the Paris Convention of 1978, and the London Dumping Convention of 1972 should be followed. Also, the International Maritime Organization (IMO) is responsible for establishing guidelines for ports to prevent and control releases and discharges from ships. (International treaties and agreements on the environment and international waterways are discussed in Chapter 2.)

Port and Harbor Location

15. Many developing countries are characterized by dense human populations, inadequate potable water and sanitary waste disposal systems, intensive land use, and increasing levels of environmental degradation. Developing a port and harbor for increasing maritime commerce and port-related industry in an area currently experiencing environmental stress may be ill-advised, unless adequate mitigative measures are planned to ensure the proper handling of wastes from development-related activities. The decision to improve or develop port or harbor facilities is usually based on economic, geographic and political parameters rather than on those of an ecologic nature. In choosing location, the assimilative capacity of the prevailing natural systems should be considered along with accessibility, employment needs and local commerce.

Project Alternatives

16. Often several project planning, design and implementation alternatives exist for development of a port and harbor facility. As a project progresses, Bank staff and consultants will describe the specific alternatives considered during project appraisal. Discussed below are various alternatives and considerations that can provide a framework for examination of a specific project by the environmental assessors and reviewers.

Site Selection

17. Selection of a site for the development of new port or harbor facilities depends on many physical characteristics of the local surroundings as well as on socioeconomic concerns. Good locations typically satisfy the following criteria.

- Physical characteristics including, wind, tides, currents, weather and siltation do not require excessive maintenance or preclude maritime traffic.
- Alterations in circulation do not position the port where maximum sedimentation will occur, leading to increased frequency of maintenance dredging.
- Shorezone land area is adequate for the processing and waste management needs of any new waterfront industries.
- Scheduling considerations such as the spawning and migration periods for indigenous biota are not violated.
- Need for resettlement is minimal.
- Project does not compete with or displace other highly valued land uses such as fishing beaches, agricultural fields or villages.

- Project activities do not adversely affect the value of an existing marine or coastal resource, such as dunes or a shellfishery.

- Construction materials, skilled labor, support industries, energy and freshwater supply, waste disposal facilities and transportation are accessible.

- Construction, operation and maintenance of the port or harbor does not damage sensitive habitats (e.g., estuaries, mangroves) or rare, threatened or endangered species.

- Port access by road/rail can be established easily without excessive disturbance to communities.

Dredged Material Disposal

18. The initial screening for evaluating disposal options is a physical and chemical analysis for geotechnical character and the presence of contaminants in the sediments. Depending on the physical and chemical character of the dredged material, disposal may be confined, unconfined, or treated prior to release in open water, along the shoreline, or on land. Disposal must be in accordance with applicable regulations. Also, long-term monitoring of the dredging process and disposal may be required.

Dredging Process

19. The primary categories of dredges include mechanical, hydraulic and innovative technologies. When selecting the appropriate dredging technology or combination of technologies, the project engineer should consider the following site-specific factors: (a) environmental constraints associated with the physical and chemical character of the sediments; (b) cost and availability of equipment; disposal site location and limitations; (c) proximity of sensitive systems such as mangroves, estuaries, regional groundwater and freshwater bodies; (d) physical conditions affecting the dredge, transit and disposal locations; and (e) interference with other users at the dredge, transit and disposal locations.

Management and Training

20. A comprehensive dredging and dredged-materials management plan should be considered for the port and harbor facilities to ensure that maintenance projects can be carried out on schedule with minimum environmental effects. Port authorities and engineers should develop a plan using data from a characterization of the materials to be dredged; a vertical and horizontal profile of contaminant distribution in the channel sediments -- to define homogeneity and locate prominent "hot spots"; an evaluation of sediment behavior using different dredging equipment and disposal options; and an analysis of potential long-term effects of maintenance on human and environmental health.

21. Support for efficient pollution control and waste reduction strategies may be important for dredging and construction activities, equipment operation, materials disposal and waterfront industries. Port and industrial engineers should be familiar with state-of-the-art equipment and disposal technologies to ensure environmentally sound waste and spoils management.

22. All project staff and laborers should receive training under "standards of practice" for occupational health and safety and emergency response. The training should include procedures to be followed in the event of accidents, spills, explosions or fires. (For detailed discussion, see the "Industrial Hazard Management" section.)

23. Training for government officials charged with supervision of an environmental management and monitoring plan may be required. To assess training needs, the capacity of local institutions to assume responsibility for environmental review should be evaluated, as well as the record of legal and regulatory agencies to monitor and enforce standards.

Monitoring

24. A site-specific environmental monitoring plan that enables Bank and local government officials to manage a project and ensure compliance with environmental standards should be prepared for each individual project. Generic parameters that may require monitoring during project planning, start-up and operation are as follows: geotechnical and chemical characterization of sediments; water quality of project area and proposed disposal area; long-term chemical/physical testing of project area; disposal location sediments and water quality; long-term monitoring of biota for the possible accumulation of contaminants; long-term monitoring of sediment transport, accretion (shoaling), erosion, and the impacts and effectiveness of any man-made structures (e.g., groins, jetties); maintenance programs to keep a high level of employee environmental awareness; and monitoring of effects of project on populations and systems in the environs of the port and harbor facilities.

Table 9.4. Port and Harbor Facilities

Potential Negative Impacts	Mitigating Measures
Direct	
1. Project location (e.g., enhancement or development of a waterway) may affect sensitive habitats and/or valuable fisheries resources or otherwise significantly degrade environmental quality.	1. Perform screening analysis of site environs and ecology and select an area that is not adjacent to sensitive habitat and would not displace valuable fisheries resources.
2. Repositioning of "null zone" in port vicinity.	2. Perform analyses to establish "controlling channel depth" which represents an equilibrium between flow-associated transport energy and sediment supply.
3. Removal and disturbance of flora and fauna at the dredging site.	3. Plan for minimizing impacts on local flora and fauna, and screen for the presence of rare, threatened or endangered species which are indigenous to the project location.
4. Interference of stationary dredging equipment with other maritime traffic.	4. Prepare a program in advance to coordinate and reduce interference to other waterway users.
5. Possible disturbance or damage to stationary installations such as underwater cables, pipelines, and outfalls.	5. • Identify and document locations of stationary installations. • Modify dredging process/disposal plans to accommodate presence of such structures.
6. Objectionable noise to nearby residents, especially at night.	6. Reduce noise level by decreasing operating level during quiescent periods in the local community.
7. Increased short-term turbidity at dredging site causing decreased light penetration and associated photosynthetic activity.	7. Reduce turbidity by efficient use of less intrusive dredging equipment, silt curtains, timing to coincide with low flow.

Table 9.4. Port and Harbor Facilities (continued)

Potential Negative Impacts	Mitigating Measures
Direct (continued)	
8. Alterations of bottom surface, which may be unfavorable to the success of indigenous benthic flora and fauna.	8. Plan for minimizing impact on important or sensitive benthic fauna and flora through ecological investigation during project planning.
9. Partitioning of natural and/or anthropogenic contaminants from sediments to the water column.	9. • Perform physical and chemical analysis of sediments prior to disturbance. • Locate potential "hot spots" and prepare plan to minimize sediment resuspension in these areas.
10. Modified bathymetry causing changes in tidal bore, river circulation, species diversity, and salinity.	10. Perform project area investigation, sampling, and characterization so project planning can yield a design which minimizes impacts.
11. Generation of turbidity plumes.	11. Use technologies such as temporary dams and/or barriers to lessen the transport of suspended material away from the project area.
12. Loss of shoreline integrity.	12. Evaluate shoreline geology and hydrology prior to project design to ensure deepening will not cause modifications such as slumping and increased erosion.
13. Upland disposal of dredged material would modify terrestrial habitat.	13. • Evaluate disposal options and select one with least impingement on important habitat. • Require reclamation plans for terrestrial sites.
14. Short-term air quality degradation resulting from dredging-related operations.	14. Monitor local air quality and reduce operations if unacceptable quality arises.

Table 9.4. Port and Harbor Facilities (continued)

Potential Negative Impacts	Mitigating Measures
Direct (continued)	
15. Projects may result in stress on local cultures.	15. • Evaluate local sociocultural environment prior to project implementation. • Develop specific mitigation measures with community involvement.
16. Covering of potential archaeological sites with dredge spoil.	16. • Evaluate disposal area for presence of important artifacts. • Modify design or salvage or protect artifacts.
17. Spills associated with increased maritime commerce.	17. • Develop spill prevention and clean-up plans. • Train a team to handle spills.
18. Covering of valuable benthic species (e.g., mussels, clams) by sediment.	18. • Monitor turbidity and maintain concentration below 2 grams/liter. • Limit dredging activity during critical spawn-and-set periods for shellfish.
19. Increasing saltwater intrusion to groundwater and surface water.	19. • Major modification to channel depth and cross section should consider the effect on saltwater encroachment. • Analysis of effects on tidal bore and river flow will be helpful.

Table 9.4. Port and Harbor Facilities (continued)

Potential Negative Impacts	Mitigating Measures
Indirect: Dredging Processes/Disposal	
20. Uptake and accumulation of resuspended and partitioned sediment contaminants by biota.	20. • Physical and chemical analyses permits proper planning prior to project implementation, thus minimizing sediment resuspension through proper selection of dredging equipment. • Implement long-term biota tissue monitoring program.
21. • Occupational health effects on workers from sediment handling operations. • Accidents occur at higher than normal frequency because of lower level of skill or labor.	21. • Train employees to be aware of potential occupational hazards and establish a facility program on safety and health which includes all of the following: • site characterization and analysis • site control • training • medical surveillance • engineering controls, work practices and personal protective equipment • monitoring and informational programs • handling raw and process materials • decontamination procedures • emergency response • illumination • sanitation at permanent and temporary facilities
22. Impacts of possible land disposal on groundwater, surface runoff and/or land use.	22. • If land disposal is the selected option, contaminated wastes should be contained in an engineered structure which minimizes leachate generation and release to local aquifers.

Table 9.4. Port and Harbor Facilities (continued)

Potential Negative Impacts	Mitigating Measures
Indirect: Dredging Processes/Disposal (continued)	
	• Uncontaminated sediments have some practical uses; however, wetlands reclamation is no longer considered an environmentally sound option.
23. Transit patterns disrupted, noise and congestion created, and pedestrian hazards aggravated by heavy trucks transporting materials to/from port and harbor facilities.	23. • Proper site selection can mitigate many of these problems, but special transportation sector studies should be prepared during project feasibility to select best routes to reduce new shorezone facilities/industries. • Develop emergency contingency plans to minimize risk of accidents during transport. • Initiate discussions on transport regulations where none exist.

LARGE-SCALE HOUSING PROJECTS

1. Large-scale housing projects are residential developments with multiple units designed as integrated schemes on single tracts of land and cover a range of development types, from subdivisions to entire cities. At the upper end of the scale the category would include the residential sectors of new towns and cities.

2. The lower end is more difficult to determine because of the ease with which housing schemes can be split into smaller phases to avoid categorization. However, 20 acres or 10 hectares are sometimes used as the lower limit of development size for compliance with certain environmental regulations, and this might be a reasonable working limit for single family detached housing, attached, courtyard, and other types of low-rise schemes, regardless of the number of units. As the densities rise, the impacts have more to do with the number of units than the size of the tract, therefore, high density row or town house schemes, low or high-rise apartments, a reasonable lower limit might be around 100 units, regardless of tract size. These lower limits are arbitrary and there should be flexibility in interpreting them. When a number of small projects begin to aggregate into a larger neighborhood, they should be treated as a large project, since experience suggests that the environmental impacts will be just as severe as a single large project, and perhaps even greater, because there may be less enforcement of regulatory standards.

3. Depending on the needs of the community, and the local socio-political framework, housing may be provided by either the public or private sectors, or a combination of both. Frequently the formal process is inadequate for community needs, and people provide their own housing using whatever materials and land is available. Most governments ultimately legitimize this squatter process by providing services and legalizing ownership. In some instances, sites and services are provided by a government agency and the units are built by individual residents. Large-scale housing projects, therefore, can come in several forms, from traditional public housing, government assisted private housing, upgrading of existing informal housing, and new sites and services projects, all of which are included in this section of the sourcebook. Because people are affected by their housing so intimately, Chapter 7 on community participation is particularly relevant to these projects.

Potential Environmental Impacts

Direct Impacts

4. Housing is an obvious and pressing human need. Until quite recently, the formidable task of providing sufficient housing for a burgeoning world population has overshadowed environmental considerations. However, with increasing pressure on land and resources has come a growing understanding of the major and severe environmental impacts generated by large-scale residential development. Many of the problems encountered in providing housing are themselves environmentally related, such as the increased costs of developing environmentally valuable, difficult, or hazardous sites. Poorly designed developments, even on essentially suitable sites, can be damaging to the environment, and threaten the health and welfare of the inhabitants. For the resident, there are many natural and man-

made conditions that have serious negative impacts on the living environment and preclude the choice of a particular site; for example, flood dangers, unstable soil conditions, seismic or volcanically activity, highly saline soils, etc. Man-made conditions are related to waste disposal areas and land that has been subjected to industrial or extractive processes, such as mining (for further discussion, see Table 9.5 at the end of this section).

5. Direct environmental impacts of housing, development occur at regional, local and site scales. The largest regional effects occur from the loss of land; prime agricultural land is often the major resource lost to development. Forests, wetlands, and habitats containing rare and endangered species, etc., are all threatened if adequate regional planning policies are not implemented. Care should therefore be taken to ensure that the long-term value of such lost or disrupted resources is identified and balanced against the need for housing.

6. Residential development contributes to air and water pollution from heating and cooling fuel use and sewage, etc. Increased solid wastes and traffic also can be expected.

7. Disruption of existing natural systems from poorly designed projects accelerate erosion and siltation, affecting both surface and groundwater quality. Groundwater quantity may be diminished due to extraction and decreased infiltration of rainwater. Increased runoff and flooding occur due to the increased impervious area (e.g., paving), and removal of vegetation and disruption of natural drainage patterns. Existing streams experience more extreme flood/drought cycles. Stormwater drainage and sanitary waste systems tax the absorption and treatment capacity of local soils and drainage networks, and groundwater becomes polluted. Erosion, subsidence, landslides, and other mechanical failure of soils and subsoils occur on improperly developed sites, particularly where there are steep slopes. Removal of vegetation may affect local climatic conditions, causing extreme temperature fluctuations and greater exposure to wind and solar radiation.

Construction Impacts

8. Building sites during construction are particularly vulnerable to environmental disturbance. Construction is often a rapid and disorderly process, with strong emphasis on completing the project and not on protecting the environment. Therefore, many seriously damaging and unnecessary impacts to the environment occur. Vegetation is removed, exposing the soil to rain, wind, and other disturbances. Excavation and grading further exacerbate this situation. Runoff increases resulting in erosion and siltation. Heavy machinery and the storage of materials compact soils making them less permeable and destroying their structure. Vegetation that is not removed may be damaged by construction equipment. Construction activity also affects the immediate surroundings of the site, e.g., by the congestion of existing roads and access points and increased noise and dirt.

Indirect Impacts

9. The manufacture, extraction, or harvesting of materials, such as bricks, cement and aggregates, coral, lumber, etc., increases during construction. This may temporarily benefit the local economy, but it may also result in shortages, the wasteful exploitation of natural resources such as forests, or the overtaxing of the local labor force. Induced development may also occur due to changes in movement

patterns caused by, for example, activities displaced by the new housing. Involuntary resettlement of existing populations may also be a factor. ("Involuntary Resettlement and "Induced Development" are discussed in Chapter 3.)

Special Issues

The Nature of Housing as a Development Type

10. While housing projects can consume considerable amounts of land they are composed of small individual units. The unitary quality of housing makes it very flexible, adaptable to almost any environment. This can be disadvantageous environmentally since it is relatively easy to build in unsuitable areas and ignore impacts and hazards at the outset. It also means that EA requires a broad knowledge of environmental impacts at many different scales. In addition, health and environmental regulations, increasing prosperity, and the use of cars as the major means of transportation, have all tended to lower project densities over the past few decades, increasing the overall consumption of land in industrial countries. Residential projects there may have influenced housing in developing countries and contributed to unnecessary environmental impacts in areas where the standards may be inappropriate for local needs and conditions.

Cultural/Environmental Relationships

11. EA of large-scale housing projects requires an understanding of indigenous lifestyles and preferences in order to achieve and maintain sustainability. New development can damage the existing cultural fabric of a region or neighborhood. A community derives its character from many generations of interaction between the people and their surroundings. Large-scale housing development introduces change not only to the natural environment, but to these living patterns, to people's relationship to the land; and the effects on the people can be significant.

12. In traditional societies, an enduring relationship between a community and the land it occupies normally indicates a relatively stable ecological balance. The relationship may not be as productive and symbiotic as in a natural ecosystem, but it is stable on a human time scale. Therefore, the extent to which a large project disrupts traditional relationships to the land should be assessed.

Mitigating Measures

Siting

13. Siting of projects to avoid ecologically sensitive, difficult, or unsafe areas is the best and most cost-effective way to minimize environmental impacts. To properly evaluate choice available, a regional database should be developed to identify and map principal environmental resources, such as major drainage patterns, freshwater and coastal wetlands, forests and other important natural habitats, prime agricultural land, etc. Where a regional database has not been developed, satellite data, aerial photography, data from academic or commercial sources, or local anecdotal information may provide a useful approximation. Such an analysis can eliminate the least suitable sites from consideration.

Site Analysis and Evaluation

14. After the initial site choice has been made, a site analysis and evaluation will identify potential environmental impacts. The goal is to understand how the site works; an extensive inventory is less important than to understand the essential nature of the site. Usually, only a few factors are critical to providing this basic overview of the site terrain, water regimen, soils and vegetation. However, a good topographic base map, with contours at about two foot intervals, is essential to site analysis and should be required on all projects.

Adaptive Strategies

15. Once the site analysis and evaluation is complete, there are several possible approaches for ensuring that a project is environmentally sound. Regulations for the protection of environmental quality are necessary, but at best, they can ensure only that minimum standards are achieved. A more effective approach is to design environmentally adaptive strategies that are built into the projects from the outset. Such strategies are derived from understanding and emulating natural models. Using permeable paving to allow percolation of water back into the soil, stabilizing steep slopes with vegetation, and treating wastewater with biological methods are examples of adaptive strategies that have been developed in recent years. Many others are possible and EAs should be designed to foster such innovative approaches.

16. This type of adaptive approach may be limited by the flexibility of local regulations, working customs, cultural expectations and affordable costs. However, experience has shown that it is invariably more cost effective than more conventional methods, once an appropriate idea is understood and accepted. For example, the preservation and use of natural drainage patterns for new projects, in lieu of piped or concrete channels with curbs and inlets, has been shown to improve flood control while lowering costs substantially. Such techniques following natural models gain economically because they require less maintenance usually.

17. Simplified site development guidelines to assist the developer and designer work with the site's natural characteristics should be written for all projects. The goal is to integrate environment awareness into the entire design of the project, minimizing the need for costly after-the-fact mitigation measures.

Design and Planning Standards

18. Evaluation of design and planning standards applied to a project may also be necessary to achieve environmentally sustainable performance, particularly if innovative adaptive strategies are to be encouraged. Building and planning standards in many developing countries were originally based on models from the industrial countries, and may be inappropriate. Standards for street widths, set backs, etc., may be overly generous, forcing consumption of too much land and increasing costs for roads and services. Reducing such requirements, provided open space set-asides are made to conserve critical environmental areas, should prove economical and environmentally beneficial as well, since less land is consumed per unit.

Management, Training, and Monitoring

19. Planners, designers, engineers, public officials in charge of project regulation and execution, developers and contractors should be educated to the environmental problems caused by many conventional development practices. Training is also required in effective site evaluation, for which an ecological perspective is needed. The person(s) conducting the EA should be familiar with the basic natural habitats of the region. After a project is complete, residents, managers, maintenance people and local officials, should be instructed in the purpose of its design and its recommended maintenance. Simplified guidelines for operation and maintenance should be distributed to encourage continuing support and understanding of the design and to build a sense of community in the new project.

20. Monitoring of environmental conditions is necessary to ensure that systems continue to operate as they were designed. Such monitoring requirements should be identified during EA, e.g., testing the quality of groundwater where wells are in use. Technical and institutional capability for monitoring should be established or strengthened as necessary, as part of the project requirements.

Agency Coordination

21. Since environment has not been of great concern to governments in the past, environmental responsibilities are often fragmented between many different agencies. The trend has been to approach this problem by forming new environmental departments (e.g., ministries) in the government. The existence of such agencies will not necessarily ensure environmental quality. They may need to be supplemented with environmental units in the implementing ministry (e.g., the ministry of housing) and at the site. Above all, effectiveness can be greatly enhanced by building grass-roots understanding and support in the affected communities and NGOs (see Chapter 7).

Table 9.5. Large-Scale Housing Projects

Potential Negative Impacts	Mitigating Measures
Direct	
1. Displacement of existing land uses.	1. • Ensure that due consideration is given to the proper trade-offs between land values for housing and those of other uses, such as prime farmland, forests, or other land uses or natural habitats of value to society as a whole. • Investigate existing planning and design standards to ensure that they are suited to local conditions and not unnecessarily wasteful of land. • Assist in drafting new regulations that are more appropriate.
2. Destruction of environmentally critical areas.	2. Ensure that regionally critical environmental sites such as major forested areas, major water bodies and wetlands, habitats containing rare and endangered species, etc., are identified and not threatened by project location.
3. Danger to residents from hazardous natural conditions.	3. • Ensure that project site is not located in the following areas: • major floodplain • coastal zone inundation areas • areas of unstable soil or subsurface conditions • areas of highly saline soils • areas subject to landslides • seismically or volcanically active areas • excessively steep or wet areas • areas where significant risk from disease vectors exist, or any other areas of significant natural hazard • Design accordingly if site cannot be moved.

Table 9.5. Large-Scale Housing Projects (continued)

Potential Negative Impacts	Mitigating Measures
Direct (continued)	
4. Danger to residents from hazardous man-made conditions.	4. • Identify areas that have significant man-made hazards such as filled land, areas subject to subsidence from mining activity, groundwater, oil or other extractive processes. • Identify areas where solid or liquid, or toxic wastes may be or have been dumped. • Investigate site conditions with proper geotechnical or chemical testing procedures. • Ensure that adequate funding and technical expertise are available to deal with the special conditions. • Investigate alternate sites.
5. Hazard to residents from air, water, or noise pollution from other adjacent or nearby land uses.	5. • Ensure that the site is located away from such pollution sources. • Do not locate down-wind of significant point sources of air pollution such as smoke stacks, for example. • Identify noise sheds around airports, major roads, etc. • Provide buffers of other compatible uses of adequate width between residential areas and sources of pollution. • Take measures to abate pollution at the source, if feasible, such as noise barriers along expressways. • Investigate alternate sites.

Table 9.5. Large-Scale Housing Projects (continued)

Potential Negative Impacts	Mitigating Measures
Direct (continued)	
6. Hazard to residents from air pollution due to site location being in an area subject to frequent temperature inversions.	6. • Seek alternate site location if pollution is from existing sources that are difficult to abate. • Otherwise, design project with low densities and non-polluting technologies for heating, cooking, etc.
7. Dislocation of existing resident populations.	7. Ensure that any involuntary resettlement is done in accordance with proper standards or consider alternate sites.
8. Destruction of historic or cultural resources.	8. • Consider alternate sites or make provision for historic culturally significant areas to be set aside in specially zoned areas. • Adapt project design to include existing historic or cultural resources.
9. Overloading of existing infrastructure and services.	9. • Coordinate with other planning goals and objectives for region. • Upgrade existing infrastructure and services, if feasible. • Consider alternate sites.
10. Excessive depletion of resources such as lumber or fuel, or overtaxing of traditional industries, such as brickmaking.	10. • Review capacity of local resources and industries to provide for large-scale construction and upgrade, if feasible. • Select materials and design criteria according to local conditions and availability of resources.

Table 9.5. Large-Scale Housing Projects (continued)

Potential Negative Impacts	Mitigating Measures
Direct (continued)	
	• Design for maximum efficiency in materials and energy use. • Encourage the study of indigenous customs and techniques for building and incorporate in project design.
Local and Site Scale Impacts	
11. Damage to sites and their immediate surroundings resulting from the disruption of the basic natural framework of the environment, in particular the soils, vegetation, and drainage network. (See below for more detailed comments.)	11. • Identify the basic natural systems of a site and its immediate surroundings and protect with set-asides for open space, easements, buffer areas, etc. • Adapt layouts to fit natural patterns rather than imposing rigid geometries.
12. Degradation of habitats caused by fragmentation.	12. Maintain and/or design open space networks to follow natural site features, such as stream corridors, and connect them throughout the site and local and regional open space systems.
13. More extreme flood/drought cycles, increased erosion and siltation and degradation of stream biota and riparian vegetation, etc., caused by increased runoff from developed sites.	13. • Preserve existing stable drainage patterns on site. • Preserve existing vegetation, particularly intact natural habitats. • Institute a stormwater management plan, which should consider such strategies as: • minimizing impervious area • increasing infiltration to soil by use of recharge areas • the use of natural vegetated swales instead of pipes, or • by using detention or retention facilities with graduated outlet control structures

Table 9.5. Large-Scale Housing Projects (continued)

Potential Negative Impacts	Mitigating Measures
Local and Site Scale Impacts (continued)	
	• Use "soft" engineering techniques for soil and bank stabilization, such as vegetative stabilization (soil bio-engineering) in preference to built structures.
14. Depletion and/or pollution of local groundwater resources.	14. • Ensure that projected use of groundwater is within capacity of natural system to replenish itself.
	• Avoid "mining" groundwater particularly in drier climates.
	• Avoid using open channel, sprinklers, or other wasteful irrigation techniques for landscaping in drier climates.
	• Use indigenous vegetation that requires less water, drip irrigation or shaded plantings.
	• Ensure that soils are suitable for septic tank or other on-site treatment.
	• Design centralized systems to avoid leakage, etc.
	• Design stormwater management systems as suggested above, in particular, use vegetation to retain, recharge and purify stormwater.
15. Degradation of soil cover from erosion, removal, or loss of soil structure due to compaction.	15. • Have both temporary (during construction) and permanent erosion control plans.

Table 9.5. Large-Scale Housing Projects (continued)

Potential Negative Impacts	Mitigating Measures
Local and Site Scale Impacts (continued)	
	• Temporary control plans should include: • silt fencing • temporary silt trap basins • short term seeding or mulching of exposed soil areas, (particularly on slopes) • limitations on access for heavy machinery and the storage of materials to avoid soil compaction
	• Permanent erosion control plans should focus on the establishment of stable native vegetation communities.
	• Ensure that topsoil in construction areas is stripped and stored for future use and not illegally removed from site.
16. Loss or degradation of vegetation from unnecessary removal or mechanical damage.	16. • Identify important stands of vegetation, large contiguous stands of forest or other natural habitat, vegetation on steep slopes, vegetated stream corridors or swales.
	• Incorporate these areas into design layout or open space system.
	• Protect such areas during construction by temporary fencing and limitations on access for heavy machinery and materials storage.

Table 9.5. Large-Scale Housing Projects (continued)

Potential Negative Impacts	Mitigating Measures
Local and Site Scale Impacts (continued)	
17. Degradation of habitat from inappropriate management or introduction of invasive exotic species.	17. • Protect natural habitat from destructive management or maintenance practices, such as the removal of understory vegetation from woodlands, or excessive clearance of vegetation from stream banks. • Do not use invasive exotic species for landscaping or reforestation.

SOLID WASTE COLLECTION AND DISPOSAL SYSTEMS

1. The overall objectives of World Bank projects for solid waste collection and disposal are to: (a) provide appropriate and efficient collection of solid wastes in the urban environment; (b) provide effective and economic transport of collected wastes to disposal facilities; (c) provide environmentally safe, technically practical and low-cost disposal; and (d) strengthen institutions technically and financially to ensure cost-effective operation and maintenance of solid waste systems over the long-term.

2. Solid waste includes: street sweepings (including dead animals); pumped sludge from septic tanks and cesspits (but not wastewater treatment plant sludge); refuse collected from residential establishments, commercial enterprises, and institutions; pharmaceutical and surgical wastes from medical clinics and hospitals; and refuse and processing wastes from industrial manufacturing facilities. The composition of solid waste varies from country to country, and culture to culture. Solid waste can contain putrescible organic matter (e.g., kitchen and market wastes, fecal matter, septage); combustible organic matter (e.g., paper, textile, and bone); and plastics, metals, glass, oil and grease, and inerts (e.g., soil and ash). Solid waste can also contain pathogenic microorganisms (e.g., bacteria and parasites), and toxic chemicals (e.g., pesticides, heavy metals, volatile organics and solvents).

3. To date, typical Bank projects in this category have included: municipal refuse collection vehicles, transfer stations and transfer trucks to increase collection service levels and lower the costs of collection and haul; septage collection vehicles and special treatment/disposal systems; workshop equipment and facilities for improved maintenance and repair of the municipal collection fleet; closure of unsanitary open dumps; refuse disposal by sanitary landfill; resource recovery by composting; pilot tests of alternative refuse collection methods for marginal zones of low-income residents and difficult access conditions; technical assistance in planning collection routes and methods; technical assistance in design and operation of disposal systems; institutional and financial strengthening of the organizational entities authorized to provide municipal solid waste management services.

4. In the future, because of need to decrease the amounts of municipal solid waste and to increase recovery of resources from them, typical Bank projects are expected to include technical assistance and policy incentives to increase segregation of secondary materials at the source and recycling; pilot tests for alternative methods of resource recovery; municipal refuse materials recycling and/or resource recovery facilities; industrial waste exchange and waste minimization.

5. Furthermore, because of the industrial growth in some developing countries, future Bank projects are expected increasingly to include special handling and disposal facilities for potentially hazardous wastes. For purposes of this document, projects which specifically deal with hazardous wastes are discussed in the "Industrial Hazard Management" section.

Potential Environmental Impacts

Public Nuisance Impacts

6. Uncollected solid waste is a public nuisance. It clogs sewers and open drains, encroaches on roadways, diminishes landscape aesthetics, and causes unpleasant odors and irritating dusts. Generally,

a solid waste project would include improving waste collection and would thus lessen the quantity of uncollected waste. However, if a project is not appropriately designed to fit in with the needs and behavioral patterns of local residents, it could result in increasing impacts related to uncollected waste.

Public Health Impacts

7. Public health can be affected when solid waste is not adequately contained at and collected from living and working environments. Furthermore, direct contact occurs when there is inadequate protection of collection and disposal workers (e.g., gloves, boots, uniforms and changing/washing facilities). As a result, the design of a solid waste project needs to consider the economic costs of waste containment and worker protection relative to potential public health impacts in order to derive an appropriate level of design.

8. Public health also can be affected when solid waste is inadequately disposed within an open dump. Although a number of solid waste projects have called for closure of open dumps and implementation of alternative sanitary disposal practices, few projects have been successful at implementing this component of the project because of land acquisition and local finance issues, as well as pressures from informal sector recycling.

9. In an open dump, there is ready access to the waste by domestic animals and, subsequently, potential spread of disease and chemical contaminants through the food chain. From an open dump, windblown dusts may carry pathogens and hazardous materials. Gases generated during biodegradation within an open dump (and to a lesser extent, a sanitary landfill) may include toxic and potentially carcinogenic volatile organics (e.g., benzene and vinyl chloride), as well as typical biodegradation by-products (e.g., methane, hydrogen sulfide, and carbon dioxide). Smoke generated from burning wastes at open dumps is a significant respiratory irritant and can cause affected populations to have a much increased susceptibility to respiratory illness.

Direct Impacts

10. Environmental damage from solid waste disposal typically can include contamination of soil, groundwater, surface water and air quality. Adverse impacts result from improper siting, inadequate design and/or poor operation. For example, seepage from solid waste contains fine particulates and microorganisms which can be filtered by soil matrices. Seepage also contains dissolved solids which can be attenuated by soil through precipitation, adsorption or ion exchange mechanisms. Under favorable hydrologic conditions, contaminated seepage (also called leachate) from solid waste can pass through the unsaturated soil beneath the solid waste deposit and enter groundwater.

11. Surface water can be contaminated as polluted groundwater is discharged into it, or by surface runoff directly from the solid waste deposit. Sources of air quality degradation include smoke from open burning; dust from inadequate containment, collection, and open dumping; and gases generated by decomposition of wastes within an open dump or sanitary landfill. (For a summary of the direct and indirect environmental impacts from solid waste collection and disposal, see Table 9.6 at the end of this section.)

Natural Resource Issues

Land Issues

12. The most obvious contamination of land is caused by windblown litter and clandestine dumping in open areas and along roadways. This contamination causes an aesthetic impact, which can result in diminished civic pride and loss of property value.

13. Soil underlying solid waste deposited within an open dump or sanitary landfill is typically contaminated by pathogenic microorganisms, heavy metals, salts and chlorinated hydrocarbons contained in seepage from the waste. The extent to which the soil attenuates such contaminants will depend on its porosity, ion exchange capacity, and ability to adsorb and precipitate dissolved solids. Furthermore, not all contaminants can be attenuated by soil. For example, anions, such as chloride and nitrate, pass readily through most soils without attenuation. Soils consisting of clay and organic matter are more likely to attenuate contaminants than soils consisting of sand, silt and gravel. If seepage continues after underlying soils have reached their full capacity to attenuate contaminants, contaminants may be released to groundwater.

14. When solid waste is processed by composting, the resulting compost product may be applied to agricultural land, wooded areas, and/or home gardens. Depending on the concentration of potentially hazardous chemicals within the compost and the land application rate used, soil can be contaminated and plants can subsequently uptake toxic chemicals. Some chemicals stay within the soil matrix and build up to phytotoxic levels after repeated applications of composting.

Water Issues

15. Through biodegradation and chemical oxidation/reduction mechanisms on deposited solid wastes, dissolved by-products of decomposition are added to the interstitial waters within the solid waste mass. Over time, the solid waste decomposes into smaller particles and the waste consolidates under its own weight, thus releasing the polluted interstitial waters.

16. Both the initial interstitial waters and any infiltration waters contaminated by decomposition by-products can seep into groundwater under certain hydrologic conditions (i.e., saturation of the waste to the point of field capacity and permeable conditions in soils underlying the wastes, as well as other hydrologic connections such as fractures in rock and inadequate casing and seals on wells).

17. Surface water can be polluted when it receives groundwater or surface runoff which has been contaminated with leachate from landfill areas. In the event that solid waste is placed in a sanitary landfill designed to enable leachate collection and leachate treatment, there may be a water quality impact attributable to the discharge of treated leachate into a receiving surface water. Potential impacts due to inadequate leachate treatment design, operational failures and bypasses, are as discussed for wastewater treatment under the category of "Wastewater Collection, Treatment, Reuse, and Disposal Systems."

Air Issues

18. The most obvious air quality problems associated with solid waste collection and disposal are dust, odors and smoke. Less obvious air quality problems may arise when the biodegradation of hazardous materials in the solid waste leads to release of potentially toxic volatile organics. For the most part, following good design and operating practices can minimize these impacts.

19. The air quality problem most associated with solid waste collection is dust created during loading operations. The level of dust created depends largely on the method of collection selected. Dust is primarily a nuisance and an eye irritant; however, it may also carry pathogenic microorganisms which could be inhaled when airborne.

20. There is typically a putrid smell from hydrogen sulfide gas and other gases created by anaerobic biodegradation of wastes within an open dump or sanitary landfill. By contrast, at a compost facility where biodegradation is designed to occur by aerobic mechanisms, the odor is typically an unoffensive earthy smell. If the compost facility is not properly operated and anaerobic conditions develop, however, a foul odor could result.

21. Burning at a disposal site may occur underground and on the surface. Once an open dump begins to burn underground, it can last for decades, or until sanitary landfill methods (including gas collection and venting) are implemented.

Social and Cultural Issues

Public Cooperation

22. In designing a solid waste collection system, social and cultural issues should be considered in order to maximize public cooperation and thus minimize costs. For guidance on involving communities in project design and implementation, see Chapter 7.

23. For example, curb-side collection can be used only in neighborhoods where residents can afford appropriate containers to be left along the curb (e.g., plastic bags or metal dustbins). When communal containers are used for collection, the distance and direction in which residents may have to walk to discharge their solid waste into a communal container should be designed to fit in with their normal routine. Where the chore of taking waste to the communal container is typically assigned to children, the height of the container or stairs up to the container opening should be designed accordingly.

Frequency of Collection

24. When neighborhoods have a high density of population and limited space available for storage of waste, waste collection frequency may need to be daily instead of every two or three days. Furthermore, when the climate is hot and humid, collection frequency needs to be daily or every two days because fly reproduction and waste decomposition rates are accelerated by heat and humidity.

Littering and Clandestine Dumping

25. Most solid waste service organizations place a high priority on providing collection service. On the other hand, they place a low priority on education and enforcement of public behavior regarding environmental regulations. The result is that the service organization unnecessarily wastes time and money trying to compensate for the uncooperative behavior of some residents by providing extra service.

26. It clearly takes more time and money (three to ten times more, is a common estimate) to pick up wastes which have been illegally discharged by littering along roadways or clandestine dumping on vacant lands. Furthermore, when the illegally discharged wastes are potentially hazardous materials (e.g., pumped sewage or industrial processing wastes), environmental impacts could be significant. Therefore, increased budgetary allocation to education, vigilance and enforcement is money well spent.

Marginal Zones

27. In the marginal zones of cities, where rural immigration and illegal settlement onto open land has occurred, providing refuse collection service is difficult. Road access for refuse collection vehicles is poor and residents may not know how to cooperate with a refuse collection system. Furthermore, where residents are settled illegally and not paying property taxes, there may be less political commitment to providing refuse collection service.

28. Based on these conditions common to marginal zones, it is typical to see clandestine dumping of wastes around the periphery of these zones, as well as on open lands between homes. As the piles of refuse accumulate, residents are likely to set fire to them at night. An understanding of the community's current practices should be sought, and grass roots efforts made to educate residents in the need for adequate refuse disposal and, to the extent possible, to help them set up relatively self-reliant refuse management systems.

Costs of Collection

29. Collection service in most developing countries consumes 30-60% of available municipal revenues. In many instances, these costs can be reduced by 30-50%. Excessive expenditure on collection service takes limited financial resources away from addressing other urban needs, such as public education. This problem can be avoided by giving adequate attention to the following in the design phase: inspection of service delivery; supervision of collection workers; selection of appropriate collection techniques; optimizing crew sizes; planning route designs; limiting direct haul to economically viable distances; and minimizing vehicle downtime for repairs.

Siting Facilities

30. In designing a solid waste disposal system, social and cultural issues arise during the siting of facilities especially. Facility siting needs to conform to land-use plans. Siting should provide for enough land area for a buffer zone to minimize aesthetic impacts. Consideration should be given to proximity to residential developments (because of noise and truck traffic impacts, as well as gas migration), prevailing wind direction (because of dust, odor, and smoke), and groundwater flow (because of water supply wells and receiving surface waters).

Recycling

31. For a significant number of the urban poor in developing countries, recovery of secondary materials is the primary source of income. Recovery of secondary materials primarily occurs as follows: informal sector workers go from door-to-door and purchase used clothing, paper, bottles, etc.; refuse collection workers sort through wastes collected while on their collection route; and pickers (also called scavengers) sort through wastes brought to land disposal sites. All such people engaged in materials recovery sell their recyclables to agents of industry. These agents provide sorting, processing, and storage to meet industry's procurement specifications. Any change in the collection or disposal system which might hinder recovery of secondary materials would have a serious impact on the feedstock and energy usage of local industry.

32. The informal sector network of recyclers is typically strongly organized despite its apparently informal nature. For example, refuse workers typically belong to a union, and pickers at disposal sites belong to either a union or cooperative. Therefore, planned changes in the collection or disposal system that would hinder materials recovery are likely to be sabotaged by the informal sector network.

Other Special Issues

Landfill Gas Migration

33. Landfill gas develops from anaerobic decomposition of wastes within a land disposal site. Unless there are competent gas control systems installed and operating at the disposal site, landfill gas can migrate underground along the paths of least resistance in the unsaturated zone (in either upgradient or downgradient directions).

34. Landfill gas can accumulate in basements of buildings along its migration pathway. Because landfill gas contains high concentrations of methane, it is potentially explosive. Landfill gas can also contain potentially toxic organic gases.

Leachate Control

35. Ideally, a solid waste landfill is located in an area where the permeability of underlying soils is very low, the nature of the soils is attenuative of dissolved chemical constituents, and uses of the receiving ground or surface waters would not be significantly affected by contamination. When less than ideal siting conditions prevail, design could include placement and compaction of a layer of relatively impermeable clay soils between the base of the landfill and the first layer of solid waste.

36. When either the nature of the waste or the site necessitates leachate collection, the issue of treatment and control must be considered. If possible, the collected leachate should be discharged to the nearest sewer to be handled as part of the area's wastewater treatment system. If no sewers are located in proximity to the landfill, on-site treatment by biological and sedimentation mechanisms should be undertaken. Recycling of the treated leachate back into the landfill system should be considered.

Medical and Toxic Wastes

37. In most cities of developing countries, there is no separate collection for medical wastes, and collection workers have no special protection for handling medical wastes and the vehicles receive no special cleaning. Medical wastes are discharged with other wastes at municipal disposal sites -- with no special means of protecting disposal site workers or pickers. Furthermore, at disposal sites where domestic animals are allowed to graze, there is the risk of reintroducing pathogenic microorganisms into the food chain.

38. To a limited extent, toxic wastes are similarly collected inadvertently as refuse collection workers service their normal routes. More commonly, however, toxic wastes are brought to municipal disposal sites by industries in their own trucks. Most disposal sites in developing countries do not have restricted access, nor do the disposal site supervisors keep any record of the nature and volume of wastes received. The wastes are dumped in the same work place as incoming refuse. Because there is no supervision of dumping, disposal workers or pickers have no forewarning of potential hazards to enable them to protect themselves. Also, there are no special safeguards at the disposal site to control the hazards that toxic wastes pose to the natural environment.

Project Alternatives

39. For various aspects of a solid waste management project, there are appropriate alternative technologies or operating methods, as listed below.

(a) **Collection Systems**

- source reduction of wastes
- self-reliant systems of on-site waste management
- equipment includes: pushcart, animal cart, tractor, and truck
- communal stationary container systems
- communal portable container systems
- curb-side collection systems from liftable containers
- block collection systems with resident cooperation
- separate collection for potentially hazardous materials

(b) **Disposal Systems**

- source reduction of wastes
- sanitary landfill (i.e., designed refuse cell construction)
- sanitary landfill with gas and leachate control
- landfill gas recovery and use
- incineration with air pollution control
- mass burn with energy recovery and air pollution control
- refuse-derived fuel production
- composting

- separate disposal zone in sanitary landfill or separate disposal site for construction/demolition debris, bulky wastes and tires
- separate disposal for potentially hazardous materials
- hold and bleed pumped sewage into wastewater treatment facilities, where available, or provide separate disposal
- separate incineration for medical wastes

(c) **Recycling Systems**

- increase product durability
- source segregation of recyclables
- manual or mechanized sorting of recyclables at transfer stations and disposal facilities
- financial incentives to private sector recycling initiatives
- refurbishing and remanufacturing of durable products
- modify procurement specifications to increase opportunities for products made from recycled materials

Management and Training

40. Solid waste management consumes a significant portion of municipal revenues. To have efficient and effective collection and disposal service, the system needs to be continuously monitored and adjusted where necessary. The managing institution, therefore, needs the appropriate authority and competence to meet these responsibilities. For example, the managing institution should be at the department level in a municipality or be set up as a public enterprise so that it could be staffed with the appropriate grade-level of professional engineers and planners. To the extent possible, the institution also needs to be empowered to generate revenues adequate to cover costs. As part of its mandate, the institution needs to be authorized to provide public education, perform inspections on public cleanliness, and enforce solid waste regulations.

41. Program support from the central government is necessary to allow local authorities to function properly, i.e., laws, regulations and policies at the central level are needed to support local ordinances, enforcement, operations and plans. Technical support from the central level may be necessary as well. Development of the country-specific state of knowledge on waste management and guidance on appropriate technology requires expertise and finance; not only are the necessary resources usually unavailable at the local level, but to develop them at other than the central level would lead to duplication of effort. For example, it is preferable for waste quantity and composition data, operational norms, service costs, available technologies, and environmental impact issues to be comparatively analyzed at the central government level, with assistance from local officials.

42. Solid waste management skills are not taught in a single curriculum in universities and adequate training seminars are not available. Solid waste managers benefit greatly by attending conferences which enable them to compare experiences. Solid waste managers would further benefit from training packages which specifically address the following: selection of appropriate collection equipment; development of collection equipment specifications; planning efficient route designs; special handling of medical wastes;

Table 9.6. Solid Waste Collection and Disposal Systems

Potential Negative Impacts	Mitigating Measures
Direct	
1. Uncollected refuse clogs open drains and sewers.	1. Provide complete refuse collection service to the urban environment.
2. Aesthetic degradation and property value loss from litter and clandestine dumping.	2. Same as above.
3. Smoke from open burning of uncollected refuse.	3. Same as above.
4. Populations of disease vectors (e.g., flies, rats, cockroaches) increase where refuse is either uncollected or open dumped.	4. Same as above.
5. Poorly located communal containers create wasted time and effort for residents.	5. Examine movement patterns of residents and survey them for the distance they would be willing to walk.
6. Lack of resident cooperation with collection systems which do not fit social and cultural behavior of residents.	6. • Survey residents for social and cultural behavior: • who performs the chore of waste discharge • at what hours are they at home • how much time could they spent on this chore • how much self-reliance will they accept • what can they afford
7. Dust during loading of refuse from stationary communal containers.	7. Minimize extra handling and maximize containment to the extent affordable.
8. Refuse scattered from stationary communal containers, plastic bags, baskets, etc., and by animals.	8. Educate residents to discharge refuse just before the scheduled time of collection service.

Table 9.6. Solid Waste Collection and Disposal Systems (continued)

Potential Negative Impacts	Mitigating Measures
Direct (continued)	
9. Worker accidents (e.g., back injuries) when refuse bins are overweight.	9. • Appropriate size refuse containers (e.g., 80 to 100 liter capacity). • Provide covers for containers so that rain does not add weight to the refuse.
10. Dust and litter along roadways used by refuse collection vehicles.	10. Provide enclosed refuse collection vehicles or cloth tarps to cover open vehicles.
11. Worker hazard when medical wastes are not specially handled.	11. • Separate collection of medical wastes using dedicated vehicles. • Provide a separate disposal area at the refuse landfill.
12. Worker hazard when potentially hazardous wastes are not specially handled.	12. • Survey industries to assess nature and quantity of hazardous wastes. • Provide separate collection and disposal in specially designed systems. • Test for waste compatibility before disposal.
13. Dust from unloading and loading operations at transfer stations.	13. Provide enclosure to the loading and unloading areas, as well as ventilation and air filtration.
14. Loss of income to pickers and loss of low-cost feedstock to industry when recovery of secondary materials is hindered.	14. • Design collection, transfer and/or disposal systems to accommodate continuation of recycling.

Table 9.6. Solid Waste Collection and Disposal Systems (continued)

Potential Negative Impacts	Mitigating Measures
Direct (continued)	
	• Increase source separation and recovery of secondary materials before waste discharge for collection. • Provide job training and employment assistance to pickers losing occupation.
15. Dust from unloading and spreading/grading operations at land disposal sites.	15. • Provide buffer zone around land disposal site. • Pave access roads. • Design location of working face to minimize truck traffic. • Water spray the working areas to suppress dust.
16. Smoke from open burning or refuse at land disposal sites.	16. Spread and compact incoming refuse, cover daily with soil, install gas control systems.
17. Odors from land disposal sites.	17. Same as above.
18. Odors from composting systems.	18. Maintain aerobic conditions during composting.
19. Contamination of ground and/or surface water by leachate from land disposal systems.	19. • Locate land disposal systems where soils underlying landfill are relatively impermeable and have attenuative properties.

Table 9.6. Solid Waste Collection and Disposal Systems (continued)

Potential Negative Impacts	Mitigating Measures
Direct (continued)	
	• Provide an adequate depth between the bottom of the landfill and the top of the groundwater. • Provide adequate horizontal distance to the nearest surface water.
20. Restriction to beneficial uses of receiving waters contaminated by leachate.	20. Do not locate land disposal upgradient of any ground or surface water whose usage could be affected by receipt of contamination, unless the distance between the land disposal and the receiving water is adequate to dilute, disperse or attenuate contamination.
21. Loss of deep-rooted vegetation (e.g., trees) from landfill gas.	21. Provide gas control systems in landfills.
22. Explosion or toxic response from accumulated landfill gas in buildings.	22. • Same as above. • Also restrict development of buildings in the vicinity of landfills. • Install gas monitoring wells around the site.
23. Emission of potentially toxic volatile organics from land disposal sites.	23. Restrict potentially hazardous wastes from being disposed in landfill for general refuse.
24. Contamination of air quality from incineration or resource recovery facilities.	24. Provide state-of-the-art air pollution control.

Table 9.6. Solid Waste Collection and Disposal Systems (continued)

Potential Negative Impacts	Mitigating Measures
Direct (continued)	
25. Contamination of soil and potential biological uptake of toxic chemicals (e.g., heavy metals) from application of compost.	25. • Based on crops planned to receive land application of compost and chemical concentrations they can tolerate. • Determine which constituent is land-limiting. • Then, based on concentrations of this constituent in compost, determine the total concentration which can be applied before phytotoxic levels are reached. • From this, determine amount of compost which can be applied.
Indirect	
26. Decline in civic pride and public morale when refuse visibly degrades the urban environment.	26. • Provide public education to obtain public cooperation with environmental regulations about littering and clandestine dumping. • Provide adequate collection and disposal service.
27. Loss of tourism when refuse visibly degrades urban environment.	27. • Same as above. • Also, provide regular street cleaning of roadways and urban environments commonly traveled by tourists.
28. Waste of municipal revenues when collection equipment is inappropriate and collection service in inefficient.	28. • Pilot test collection systems before implementation. • Regularly evaluate costs of collection in various neighborhoods by various techniques. • Undertake measures to lower costs and improve service.

Table 9.6. Solid Waste Collection and Disposal Systems (continued)

Potential Negative Impacts	Mitigating Measures
Indirect (continued)	
	• Regularly design optimum routes for collection.
	• Implement transfer systems whenever direct haul is uneconomic (e.g., when travel times are over half an hour or more than 15 km).
	• Increase inspection and supervision to obtain improved productivity and effectiveness by collection workers.
	• Decrease vehicle downtime by adequate planning of the spare parts inventory needs.
29. **Land use conflicts when solid waste facilities are not well located.**	29. • Plan locations of facilities to fit existing and planned land use.
	• Provide buffer zones to minimize aesthetic impact of facilities.
	• Try to limit truck traffic to use of major roadways which do not have residential development.
30. **Public opposition to proposed solid waste facilities.**	30. • Provide adequate operation of existing facilities.
	• Provide a demonstration of how the proposed facility will be operated.
	• Perform public relations early in the facility planning process, including visual aids showing similar facilities in other places.

Table 9.6. Solid Waste Collection and Disposal Systems (continued)

Potential Negative Impacts	Mitigating Measures
Indirect (continued)	
31. Deterrent to industrial development by environmentally conscious industries when competent disposal facilities do not exist for hazardous wastes.	31. • Provide special facilities for receipt of potentially hazardous wastes. • Provide equitable environmental regulation and enforcement at the national level so that all industries are operating under the same environmental standards.
32. Loss in public's faith in the political system when inappropriate solid waste facilities (e.g., incinerators) are constructed and not used.	32. • Set up institutional arrangements, such as an interministerial task force whose mandate is to review technical and economic viability of large projects and prioritize them for financial assistance (as in Indonesia). • Set up institutional mechanisms whereby all large projects have an environmental and economic impact assessment presented for review at the national level prior to receiving approval for implementation (as in Thailand).
33. Increase in country trade imbalance and energy consumption when there is a decrease in recycling of secondary materials from wastes into industrial feedstock.	33. • Provide incentives to private sector entrepreneurial initiatives in recovering secondary materials or recycling. • Improve government procurement specifications so that producing products from recycled materials in encouraged. • Provide public education which encourages recycling. • Facilitate source segregation of recyclables and separate collection. • Design transfer systems and/or disposal systems to accommodate recycling from mixed refuse.

TOURISM DEVELOPMENT

1. Tourism contributes significantly to the economies of developing countries. Growth in the sector has been more rapid there than in developed countries and has been continuous for several decades. Tourism projects may involve site identification and provision of access; construction of hotels and other visitor accommodations and amenities; creation of duty-free zones; and the establishment of facilities such as sports complexes, marinas and parks for other leisure-time activities. Supporting infrastructure is an important component of any tourism project. Park facilities, roads, solid waste collection and disposal, sewerage and drainage, and water distribution systems often need to be constructed or improved. Institutional strengthening is often funded under this category.

Potential Environmental Impacts

2. Tourism projects frequently are comparatively small and, when screened for potential environmental impacts, are often placed in Category B. They warrant attention in the Sourcebook nevertheless: first, because of the close relationship between tourism and environmental quality; and second, because of the many linkages between tourism development and other sectors in the same region.

3. The features of the natural and sociocultural environment that are important resources for tourism attract people because of aesthetic, recreational or educational/scientific value. However, many of the same features are particularly sensitive to disturbance by human activities. Negative impacts resulting from inadequately planned and uncontrolled tourism development can easily damage the very environments on which the success of the project depended (see Table 9.7 at the end of this section for examples). This in turn may severely reduce project benefits. In other words, without careful attention to the balance between the volume and type of tourist activity and the sensitivities and carrying capacities of the resources being developed, tourism projects can be not only environmentally harmful but also economically self-defeating. For example, an increasing number of hotels sited to attract tourists to a coral reef fail after a few years, because hotel effluents discharged offshore rapidly impair -- or kill -- the reef.

4. Tourists increase demands on local infrastructure -- transportation, water supply, wastewater collection and treatment, solid waste disposal, and health care facilities -- and on the variety of public services that are usually the responsibility of local government. Often the demands have significant seasonal peaks. Without coordination and planning, service demands may exceed capacity, with adverse results for residents as well as tourists.

5. Indirect linkages between tourism and local cultures, businesses, resident populations and workforces are potential problems. Failure to recognize them can diminish project benefits as well as inflict adverse socioeconomic impacts on the local population. For example, commercialization of traditional artisan industries can lead to loss of authenticity with negative results for the artisans and possibly for the buyers as well.

6. The magnitude and scale of impacts depend on the size and type of tourism development proposed, relative to the fragility of its proposed environment. Recreational tourism involving a variety of sporting activities and a large hotel complex infrastructure has a greater potential to degrade fragile ecosystems than projects which attempt to attract tourists with scientific or educational interests such as birding, nature photography or archaeology.

7. On the positive side, "ecotourism" projects can combine conservaiton of natural and cultural sites with economic and recreational benefits. Success depends on informed site selection, sound design and operating guidelines which take into account the sensitivity and capacity of the resources which form the tourist attraction. Consequently, a major concern in planning other types of development and analyzing their impacts is to avoid foreclosing tourism development options by degrading resources especially well-suited to it. Comprehensive environmental and land-use planning can identify options and alternatives over the long term and balance single and multiple use concepts.

Special Issues

8. Availability of clean water for drinking, provision of wastewater treatment consistent with the capacity of local water bodies to assimilate pollution load, and adequate facilities for solid waste disposal are critical issues for this sector. If these services are provided by local government or independent utilities, the project sponsor should demonstrate that detailed information on the tourism development has been furnished to those agencies and that they are prepared and able to meet the project's needs. If the services are not available from local agencies, the plan for the project should show clearly how the developer proposes to provide them, and the impacts of the proposal should be considered in any EA or other environmental analysis. In either case, planners should recognize that tourists from industrial countries use more water and other resources and generate more waste per capita than do residents of developing countries.

9. Coastal zones are among the most attractive areas to tourists. Consequently, tourism constitutes an additional development pressure in areas already heavily used for ports and harbors, commercial fisheries and shell fisheries, and urban expansion. Too, the tendency for developers to seek out new, "unspoiled" sites, away from already congested beaches and towns, contributes to the trend toward urbanization of entire coastlines. The environmental effects of seasonal population increases can be particularly significant along the coast (see para 11). For additional discussion of coastal and marine resources, see the "Coastal Zone Management" section in Chapter 2.

10. Most islands tend to be environmentally fragile and highly vulnerable to development pressures, especially from tourism. They tend to have distinctive flora and fauna which can be displaced by non-native species, which may be introduced during development. Island populations often include indigenous peoples with distinctive cultures. Natural resources such as coral reefs, seagrass beds, mangroves, tropical forests, waterfalls, caves, gorges, and geothermal areas are frequently concentrated in small areas which are attractive for tourism. Fisheries and other subsistence economies prevalent on islands can undergo transformation with even modest scale tourism development. The resource base of islands over the last decade shows major decreases in environmental quality from increased sediment, pesticide loading, eutrophication from fertilizers and sewage, coastal development, and industrial effluent discharge.

11. Because of the seasonal nature of many tourist activities, demands at peak periods may exceed the capacities of local public services and physical infrastructure. Typical problems are traffic congestion and demands in excess of capacity of water supply, wastewater and solid waste disposal systems. Wildlife may be affected by large influxes of people at the critical times of migration, feeding, breeding or nesting.

12. Sociocultural considerations are particularly important in environmental assessment of tourism projects. Activities such as tours of archaeological sites may conflict with local religious beliefs. Hotel construction may cause displacement and involuntary resettlement. Induced development may occur at the fringes of tourist areas. The influx of large numbers of foreigners (tourists or migrant workers) into a local culture and the likely clash of contrasting life styles that result, can have serious impacts on local cultures. There is also the risk of exploitation of indigenous cultures, music and folklore.

13. The visual as well as the physical impact of accommodations and other structures that will be built to serve tourists should be considered. Ease of construction and "efficient" design should be tempered by considerations for harmony with the surrounding natural environment and sociocultural context. The impact of tourism infrastructure on resources valued for their aesthetics (e.g., waterfalls, gorges) view should be specifically addressed. Also, tariffs for water, sewerage, and other services may be necessary to avoid burdening local users unfairly.

14. Assessments of tourism projects should include analysis of the projected distribution of costs and benefits. Whereas the benefits of tourism may be assumed to accrue to local residents, residents are likely to incur more of the costs and may enjoy less of the benefits than visitors, immigrant workers or commercial intermediaries. For example, if high-quality employment opportunities are expected to result, how many jobs will be made available to local residents and for how long, especially if training is required to qualify them for the work? National or regional laws and regulations concerning expatriate employment will provide a base for evaluation of probable impacts.

Project Alternatives

15. Environmental assessment incorporates the concept of alternatives to the proposed project or to the ways of executing it. During project planning, alternatives should be identified and described in ecological, technical, economic and social terms for decision makers. Special impacts associated with each alternative, suitability under local conditions, and institutional, training and environmental monitoring requirements will be identified and compared to the resources available.

16. A number of plans and strategies may be required to implement a sound tourism project. At minimum, a land-use management plan and a pollution control plan would enable environmental objectives to be incorporated early into the development process. Integrated planning is particularly desirable for tourism projects.

Scale of Development

17. Alternatives may address the overall scale of development appropriate for the region. At one end of the range is small-scale, low-impact tourism -- e.g., a wildlife refuge or field research station with low numbers of visitors (fewer than 100 at any time), offering forms of recreation such as hiking, nature photography and birding, and oriented toward education of the tourist. At the other end is large scale, high impact tourism, including major infrastructure development with all amenities, capacity for thousands of visitors, high density accommodations, and a wide range of recreation. Initial planning should consider scale and carrying capacity.

Siting

18. Facility siting is a principal factor, not only for the process of the tourist development but for any attending adverse impacts, including population displacement and degradation or loss of natural and cultural resources. Frequently, unique habitats, natural hazards, beach or soil erosion, saltwater intrusion and other natural processes are not properly characterized, resulting in selection of unsuitable sites. Location should not be based on the "best looking" beach or most majestic scenic view but should be the result of a critical examination of alternative sites.

Management and Training

19. Institutional support may be required to enable the tourism development project to succeed. Investments may need to consider public expenditures for infrastructure and commitments of staff and equipment beyond the means of local institutions to respond and manage the new development. If the country, region or locale has had a significant influx of tourism in the past there may be a small level of added support required to focus better on the environmental aspects of the project. Underdeveloped or undeveloped areas may require management and training in the management for the local natural resources staff (e.g., parks, fisheries, forestry officials), roads and sewer authorities, local pollution control authorities and environmental protection agencies, and even the institution responsible for tourism. The local labor force may need training in order to compete for jobs generated by the project and thus to participate fully in its benefits.

20. Frequently, legislative actions involving wildlife protection; national trusts such as parks, historic buildings, and archeological sites; accommodation of land, sea and water rights of indigenous peoples; and general environmental management of sensitive habitats are needed. Tourism development requires coordinated management between agencies responsible for tourism, parks and reserves, and pollution control in order to avoid the deterioration of environmental resources. Management training, legal assistance, and organizational restructuring may be required: (a) to establish, monitor and enforce vironmental legislation and standards; (b) to investigate, plan, and monitor potential adverse effects of pollution; (c) to mitigate and control such pollution or other adverse impacts resulting from tourism; (d) to provide assistance to nongovernmental organizations and others working to prevent deterioration of natural and cultural resources in a proposed development area; and (e) to assess social effects on local communities and reduce or plan human resettlement.

Monitoring

21. Monitoring plans should include baseline data and periodic review of objectives to determine if plans are being realized. Typical profiles can be developed for protected and ecologically sensitive areas such as beaches, wetlands, reefs; water quality and sediment loading in all water bodies; erosion and sedimentation impacts associated with infrastructure development such as roads, ports, harbors, marinas, hotels, shopping centers and the like; impacts associated with recreational activities such as reef diving, spear fishing, use of all-terrain vehicles, and access to areas previously denied; degree of staging/phasing of development and any observed impacts; demands on transportation and other infrastructure such as water supply, wastewater treatment and solid waste disposal capacity, and the observed system responses; effects on local and regional society and economy.

WATER SUPPLY

1. Water supply projects involve any or all of the following: construction, expansion or rehabilitation of dams and reservoirs, wells and intake structures, transmission mains and pumping stations, and treatment works and distribution systems; provisions for operation and maintenance of any of the facilities just listed; establishment or strengthening of metering, billing and collection functions; and strengthening of overall water utility management. This edition of the Sourcebook does not contain a single detailed section on water supply projects. However, most of the environmental issues which pertain to them are presented in the chapters listed below:

Chapter 2: Land and Water Resource Management

Chapter 8: Dams and Reservoirs

Chapter 9: Wastewater Collection, Treatment, Reuse, and Disposal Systems

Chapter 10: Oil and Gas Pipelines

Table 9.7. Tourism Development

Potential Negative Impacts	Mitigating Measures
Direct	
1. • Beach mining of sand for construction.	1. • Control of construction contractor.
• Destruction of reef for aggregate materials used in construction.	• Submission of plans in accordance with local ordinances on beach sand mining.
2. Destruction of wetlands, forests, other unique/sensitive habitats or cultural, historical and archaeologically important sites.	2. • Areas considered for development should have zoning plans to account for natural geographic and socioeconomic condition.
	• Base development phase on an inventory of resources.
3. Erosion resulting from uncontrolled clearing, infrastructure construction such as roads and marinas.	3. Develop erosion and sediment control plans.
4. Loss of "free" environmental services from natural systems and degradation of air, water, land resources.	4. • Carrying capacity should be defined so that target tourist population can be sustained without overburdening existing infrastructure and resources.
	• Include improvements in project design.
5. • Water pollution from inappropriate sewage or solid waste disposal. • marine effluent disposal • residential sewage disposal • marinas • infiltration to groundwater	5. • Allowance made for use of existing municipal or regional collection and disposal system or construction of on-site sewage treatment plant.
	• Liquid waste should not be discharged onto beaches, coral reefs, or other sensitive areas.
	• Verify local capacity to monitor and enforce pollution regulations.

Table 9.7. Tourism Development (continued)

Potential Negative Impacts	Mitigating Measures
Direct (continued)	
6. Solid and liquid waste disposal creates nuisance conditions adjacent to amenities.	6. • Appropriate waste disposal options required to manage potential problem. • Landfill versus incineration alternatives, as well as waste minimization will be considered.
7. • Access problems created: • traffic congestion • noise • minor and localized air pollution • people density greater than services available	7. Access problems minimized by integrated planning to reduce traffic and pedestrian congestion, noise.
8. Sea turtle nesting affected (special case).	8. • Beach monitoring for turtle protection coupled with beach zoning and development guideline to preserve the natural beach environment from the primary dune seaward. • Restricting night activities on nesting beaches during egg-laying and incubation periods.
9. Displacement of human population.	9. • Plan and implement program of compensation and resettlement. • See Chapter 3 for discussion of involuntary resettlement concerns.

Table 9.7. Tourism Development (continued)

Potential Negative Impacts	Mitigating Measures
Indirect	
1. Conflicts with other resource use such as fisheries, agriculture.	1. • Conceive tourism development in framework of national, regional, local socioeconomic development plans to integrate new objectives into development strategies. • Identify zones most suitable for tourism.
2. • Stress to capacity to manage the "tourist or related environment." • legislation and polling constraints • agency support lacking • staffing and financial resources to mitigate impacts absent/reduced • inadequate training in environmental management	2. • Comprehensive legislative action frequently required to address direct and indirect impacts and their monitoring and evaluation. • Staffing and equipment support must be budgeted, including whatever training necessary to mitigate impacts and monitor the "environmental protection plan" or other mitigation plan.
3. Multiplier effect on other industries causes increased stress on natural resources or services (craft market, vendor, taxi driver, suppliers, farmers/fishermen).	3. • Provide adequate infrastructure and services support to meet physical, social and economic needs of the region. • Recognize that "overbuilding" may be a persistent problem.
4. Congestion, overcrowding.	4. Design (urban areas and transport networks, etc.) according to carrying capacity of natural setting.
5. Natural hazards peculiar to developed site such as coastal storms, flooding, landslides, earthquakes, hurricanes, volcanos, may stress infrastructure and reduce long-term benefits.	5. Design facilities to: (a) meet best possible specifications for natural hazard amelioration; (b) take advantage of natural resources such as wetlands ability to buffer storms or absorb treated wastewater (see "Natural Hazards" section).

WASTEWATER COLLECTION, TREATMENT, REUSE, AND DISPOSAL SYSTEMS

1. This category includes new projects and project components and rehabilitation activities involving: conduits for collection and conveyance of wastewater, pumping stations, conventional and innovative treatment works, wastewater reclamation and reuse projects, ocean outfalls, wastewater treatment sludge management facilities, a variety of small-scale sanitation systems for rural and urban areas, and urban stormwater drainage projects. Where water quality problems are severe, as is the case in many densely-populated urban areas, individual wastewater projects may be executed as increments of long-term pollution control programs whose ultimate objectives realistically may take 10 to 20 years or more to achieve. Water pollution control programs often include significant institution-building and national water pollution control policy formulation components.

Potential Environmental Impacts

2. The pollutants in municipal wastewater are suspended and dissolved solids consisting of inorganic and organic matter, nutrients, oil and grease, toxic substances, and pathogenic microorganisms. Urban stormwater can contain the same pollutants, sometimes in surprisingly high concentrations. Human wastes that are not properly treated and are disposed of at the point of origin or are collected and carried away pose risks of parasitic infections (through direct contact with fecal material) and hepatitis and various gastrointestinal diseases including cholera and typhoid (through contamination of water supplies and food).

3. When wastewater is collected but not treated properly before disposal or reuse, the same public health hazards exist at the point of discharge. If such discharge is to receiving water, additional harmful effects will occur (e.g., habitat for aquatic and marine life is impaired by accumulated solids; oxygen is depleted by decomposition of organic material; and aquatic and marine organisms may be further harmed by toxic substances, which may spread to higher organisms through bioaccumulation in food chains). If the discharge enters confined waters such as a lake or bay, its nutrient content can cause eutrophication, with nuisance plant growth which can disrupt fisheries and recreation. Solid waste generated in wastewater treatment (grit, screenings and primary and secondary sludge) can pollute soil and groundwater if not properly handled.

4. Wastewater projects are executed in order to prevent or alleviate the effects of the pollutants described above on the human and natural environments. When properly carried out, their overall environmental impact is positive. Direct impacts include abatement of nuisances and public health hazards in the serviced area, improvement in receiving water quality, and increases in the beneficial uses of receiving waters. In addition, installation of a wastewater collection and treatment system provides an opportunity for more effective control of industrial wastewater through pretreatment and connection to public sewers and offers the potential for beneficial reuse of treated effluent and sludge. Indirect impacts include the provision of serviced sites for development, increased fishery productivity and revenues, increased tourist and recreational activity and revenues, increased agricultural and silvicultural

productivity and/or reduced chemical fertilizer requirements if treated effluent and sludge are reused, and reduced demands on other water sources as a result of effluent reuse.

5. A number of these potential positive impacts lend themselves to measurement and thus can be incorporated quantitatively into analyses of the costs and benefits of various alternatives when planning wastewater projects. Human health benefits can be measured, for example, by estimating avoided costs in the form of health care expenditures and lost workdays which would result from poor sanitation. Reduced drinking and industrial water treatment costs and increased fishery, tourism and recreation revenues can serve as partial measures of the benefits of improved receiving water quality. In a region where demand for housing is high, the benefits of providing serviced lots may be reflected in part by the cost differential between installing the infrastructure in advance or retrofitting unplanned communities.

6. Systems in which treated wastewater or sludge are reused may be more expensive to construct and operate than those in which the sludge is disposed of as a waste product. In evaluating alternatives involving reuse, however, it is important to include such benefits as increased water availability to support development in the region, the opportunity to diminish irrigation demands on potential public water supply sources, reduced need for chemical fertilizers, incremental improvements in crop and timber production, and low-cost means to re-vegetate marginal soils or reclaim them for agriculture or silviculture. These too can often be measured, most of them by calculating avoided costs. Chapter 4 provides more details on economic evaluation methods.

7. Unless they are correctly planned, sited, designed, constructed, operated and maintained, wastewater projects are likely to have a negative impact overall, failing to yield the full benefits for which the investment was made and adversely affecting other aspects of the environment besides. The individual items listed are self-explanatory, for the most part, and will not be discussed in detail in the text. However, there are several characteristics common to many of the potential impacts and mitigating measures which should be emphasized as special issues throughout project preparation, assessment and implementation. These are: first, the importance of sound and comprehensive wastewater system planning; second, the fundamental dependence of wastewater projects on proper operation and maintenance (and thus on strong institutional support for both); third, selection of appropriate technology; fourth, the necessity for an effective industrial wastewater pretreatment program in any municipal system serving industrial customers; and finally, the need to consider a number of potential sociocultural impacts which are sometimes ignored in project preparation (see Table 9.8 at the end of this section for a summary of all potential impacts and recommended mitigating measures).

Special Issues

Planning

8. Decisions on the type and location of effluent discharge and the level of treatment are crucial and should not be made without adequate information. Present wastewater volume and strength are basic information for the planning process, and realistic projections of volume are important to establish the magnitude and timing of the need for collection and treatment. In making and updating projections, other planned development activities should be taken into account so that extensions or expansions of wastewater infrastructure can be coordinated with them.

9. Level of treatment -- the magnitude of pollutant removal a treatment process must achieve -- depends on the performance standards which apply to the system. These are usually expressed as limitations on the concentrations of regulated substances permitted in the treated effluent. In the case of effluents which are to be applied to crops or otherwise used on land, the standards are set to prevent crop and groundwater contamination. National standards for effluent reuse may already exist; if they do not, they can be based on World Health Organization or World Bank guidelines or derived from the standards of other countries in which land application practiced.

10. For discharges to surface waters, the process of setting standards often begins with classification of receiving waters based on the intended or desired uses. In the water quality standards of the People's Republic of China (PRC), for example, there are five classes ranging from Class I "very clean natural waters" to Class V "suitable for cooling water only in industry and can be used for agriculture." Receiving water quality standards can then be established on the basis of the scientific literature to provide for the uses in each classification. In the PRC, 29 parameters are covered for each surface water class. Classification of waters should be done with consideration for what is economically and technically realistic; requiring drinking water quality in a busy harbor, for instance, is not a sound use of pollution control resources.

11. Ideally, effluent limitations for wastewater dischargers should be determined by a mathematical modelling process which takes into account the existing quality and flow characteristics of the receiving water body, calculates the maximum load of each pollutant which can be assimilated in each segment or zone under a specified statistical condition of dry-season streamflow (e.g., the minimum monthly flow in a five-year period) without causing the standards to be violated, and allocates that load among all dischargers. Such models require seasonal data on receiving water quality, volume and concentration of all discharges, and a long enough record of hydrologic data to show seasonal flow averages and permit calculation of the dry-season flow.

12. In practice, national effluent limitations are often established to correspond to the various receiving water classes, to simplify the process of preparing discharge permits or establishing base/minimum levels of treatment. Modelling is reserved for situations in which adherence to those limits will not result in attainment of water quality standards and more stringent requirements must be applied (or where projects are being planned in countries without water quality or discharge regulations). PRC regulations prohibit any pollutant discharge to Class I and II waters and provide two sets of effluent limits for other waters -- one for Classes III and IV and another, less restrictive, for Class V. PRC law specifies quantitative procedures whereby local discharge limits more strict than national standards may be set where necessary to achieve desired uses.

13. Limitations on discharges to marine waters are usually simpler; they focus on preventing discoloration of the water and pollution by oil and grease, floating debris, and bacteria (in shellfish harvesting and recreational waters). The major planning task is to identify an acceptable location for the submerged outfall, where the effluent will not degrade significant water areas or contaminate shellfish beds and beaches. There are mathematical models for this purpose, which simulate the processes of dilution, dispersion, diffusion, stratification and pollutant decay or die-off. The models require current, temperature, salinity and water quality data collected over a full 12-month cycle, along with detailed bathymetric and ecological information.

14. Data collection and modelling are costly and time-consuming. However, in comparison to the capital and operating costs and expected life of wastewater systems, both the costs and the time are inconsequential, and the benefits -- facilities which neither fall short of their water quality objective nor involve unnecessary expenditures -- are substantial. Furthermore, data collection can sometimes occur in parallel with design and construction activities when wastewater projects are phased as described below.

15. A second component of planning involves the sequencing and phasing of projects within individual wastewater systems, as parts of long-term pollution abatement programs, and in relation to activities in other sectors. For example, putting a collection system into operation without the treatment works merely concentrates harmful discharge and is a frequent cause of gross pollution of surface waters. The net effect is to increase water scarcity or increase the cost of treating water for drinking or other uses. Installing public water supply and proceeding with residential, commercial or tourist development will cause public health hazard or water pollution if wastewater infrastructure is not provided at the same time.

16. In many cases, it is cost-effective to construct treatment works in a modular fashion, adding additional capacity as the collector system is extended and new connections are made. Phasing wastewater investments may be the only realistic way to make progress toward ultimate water quality objectives in densely populated, highly polluted areas, where a single project would exhaust all resources available for public works and physically disrupt the region. The level of treatment can be phased in a single project or as part of sector strategy, an approach which is helpful when environmental improvement is urgently needed but local financial resources are limited or the scientific data to determine exactly the extent of pollutant removal required has not been collected. It is important in any phased approach to reserve space for future expansion when acquiring sites and designing facilities.

17. Wastewater treatment generates sludge and other solid waste, such as grit, and grease screenings. Finding locations for landfill or incineration, or outlets for recycling, is often difficult. However, if solutions are not found, a portion of the pollutants removed from the wastewater will become pollutants of the land. Sludge management should be part of wastewater system planning.

Operation and Maintenance

18. Treatment works and pumping stations will not perform properly unless they are operated and maintained properly. The most common causes of system failure are selection of inappropriate technology, lack of spare parts, shortages of trained operators and skilled technicians and craftsmen, and unreliable power or chemical supplies. Most of these in turn can be traced to institutional weaknesses in technical training and utility management, inadequate operating budgets and unattractive salaries (both often related to artificially low customer charges for water which do not produce revenues to cover the full cost of water supply and wastewater disposal services).

Selection of Appropriate Technology

19. The concept of appropriate technology in wastewater systems has technical, institutional, social and economic dimensions. From a technical and institutional standpoint, selection of inappropriate technology has already been named as a primary cause of system failure. The wastewater environment is a hostile one for electronic, electrical and mechanical equipment. Maintenance is a never-ending

process, and it requires support -- spare parts, laboratories, trained technicians, specialized technical assistance, and adequate budgets. Even in developed countries, it is the simpler systems, selected and designed with maintenance in mind, which provide more reliable service. In developing countries, where some of the ingredients for a successful maintenance program may be lacking, this should be the first consideration in choosing treatment plant and pumping station technology.

20. In small communities and rural settings, the technical options tend to be simpler, but the institutional considerations blend with the social and remain extremely important. Local institutions must be able to manage the sanitation programs or systems. Community participation may be a key element in their success. Social preferences and customary practices are significant; some can be modified by education programs, but others may be rooted in cultural values and not subject to change. Economics become part of the decision in two ways. Not surprisingly, the simpler technologies, selected for ease of operation and maintenance, tend to be less expensive to construct and operate. Even when they are not, however, as may be the case when a great deal of land has to be acquired for stabilization ponds, a less expensive system which fails will ultimately be more costly than a more expensive one which operates reliably.

Industrial Wastewater

21. In most circumstances, connecting industries to public sewer systems is sensible. It reduces the number of discharge points and thus the complexity and cost of monitoring and enforcement, provides the opportunity for better control of industrial effluent, and may be lower in total cost. Absolutely critical to success, however, is an industrial pretreatment program including regulations with specific limits on discharges of hazardous and toxic substances and other pollutants to public sewers, monitoring procedures, and enforcement capability. Otherwise, there is a risk of exposing water system personnel and components to hazardous materials, disrupting the treatment process, transporting toxic pollutants to receiving waters or the land, and contaminating treatment plant sludge so severely it cannot be put to beneficial use or even disposed of without difficulty.

Sociocultural Issues

22. Treatment facilities require land; siting them can lead to involuntary resettlement. Moreover, treatment and disposal works can cause nuisances in the immediate vicinity, at least occasionally. Often, the lands and neighborhoods selected are those of "vulnerable groups" who can least afford the costs of dislocation and whose living environment is already impaired. Care should be taken to site treatment and disposal facilities where odors or noise will not disturb residents or other users of the area, to manage resettlement with sensitivity (see "Involuntary Resettlement" section in Chapter 2), and to include in the project mitigation plan provisions to mitigate or offset adverse impacts on the human environment. If these considerations are not included in project planning, there is substantial risk of solving one community's environmental problem by transferring it to another.

Project Alternatives

23. A variety of siting and technological alternatives exists for wastewater collection, treatment and disposal, and sludge management. Several will be applicable in every situation.

(a) Collection Systems

- on-lot treatment
- individual holding tanks with truck collection
- small-diameter gravity, pressure or vacuum sewers
- shallow sewers
- flat sewers
- simplified sewerage systems
- conventional gravity sewers and force mains
- regional collection systems
- community or sub-regional systems

(b) Treatment Works

- community on-lot systems
- oxidation ditches
- stabilization ponds
- aerated lagoons
- artificial wetlands (or constructed wetlands)
- land treatment
- conventional biological treatment
- physical-chemical treatment
- preliminary or primary treatment with ocean disposal

(c) Disposal

- reuse in agriculture, silviculture, aquaculture, landscaping
- reuse for groundwater recharge
- rapid infiltration
- underground injection
- reuse in industrial applications
- ocean outfall
- surface water discharge
- nightsoil treatment plants

(d) Sludge Management

- composting
- co-composting with municipal refuse
- reuse in agriculture or silviculture
- reclamation of marginal land for reforestation, cultivation
- energy recovery (methanization)
- incineration
- landfill
- ocean disposal

Management and Training

24. Institutional support is critical to successful wastewater utility operation. It begins with staffing. A large collection and treatment system will need a manager with technical and administrative experience; an environmental engineer; supervisors and workers in operations, sewer maintenance and plant maintenance crafts; laboratory directors and technicians; and support staff in accounting, budgeting, and clerical areas. If the project includes on-site disposal systems, holding tanks or small-diameter sewers with settling tanks, staff will be needed to develop and enforce standards for the facilities, and to inspect and approve installations, and there will have to be provision for on-site system maintenance. A customer relations unit is essential to receive and investigate customer complaints, provide information to customers, and conduct education programs related to the system's services (e.g., hygiene and sanitation, on-site system upkeep). If the utility is responsible for revenue collection, a billing and collections group will be required. Job status and salaries must be adequate to attract qualified personnel and to avoid losing them once they have been trained.

25. Wastewater systems can be operated successfully under a variety of organizational structures such as a department of municipal government, a local or regional sanitary district or authority, or a private operating company under contract to the government. For conventional sewer systems, the planning, construction, operation and administrative functions can be centralized within one of these organizational formats. For projects which involve individual on-site systems, or small community equivalents to them, some decentralization of function is appropriate. Planning the areas to be served by various types of on-site systems and setting standards for construction, installation, and maintenance should remain more or less centralized.

26. Permitting and inspecting installations and enforcing standards should be a governmental function; it is logical to delegate it to local authorities but it could be carried out by a centralized utility, in situations where the wastewater system is a hybrid of on-site and conventional sewerage. Technical assistance with installation and education of users in operation and maintenance are services best performed at the community level, by local authorities or NGOs. System maintenance, including septic tank pumping, is a service which should probably be privatized and carried out under supervision of local authorities. Where companies do not exist to perform maintenance reliably, having a centralized wastewater utility undertake it is an option. Septage disposal should be regulated according to government standards and supervised by the agency which administers the disposal facility, site, or reuse program.

27. Training should begin before system start-up, with the assistance of the design consultant. Besides basic familiarization with the system, its relationship to the environment, and fundamentals of occupational health and safety, it should include operations and maintenance training on the actual equipment. Industrial waste control staff will need specialized training in sampling and enforcement procedures. Employees of connected industry will need training on operation and maintenance of pretreatment equipment. Any personnel who may have to enter confined spaces, work in deep trenches, maintain electrical equipment, or handle chlorine or other dangerous chemicals must be properly equipped and drilled in safety and accident response procedures. "Training the trainers" is a good concept, so that system staff can continue training programs throughout the life of the facilities.

28. A wastewater utility needs to recover its full costs of service to sustain reliable operations. It thus must have or be able to obtain the capability to determine costs and develop rates. Rates typically include a fixed charge based on capital cost recovery and system administration, plus a variable charge based on a customer's metered water consumption. Surcharges are an appropriate means to recover the additional cost of serving industrial customers with discharges significantly exceeding the average strength of domestic wastewater. Where wastewater services are provided at less than full cost for some groups of customers as a matter of government policy, the government should make necessary provisions to avoid operating budget shortfalls, perhaps through intra-urban cross-subsidies.

29. A start-up plan should be prepared for any new wastewater facility of significant size to ensure that the requirements described above are met. The plan should provide for assembling staff, maintenance equipment and spare parts in advance of need, training all personnel, and establishing revenue sources and budget.

30. Local, regional and/or national public health and environmental agencies will need the resources and training to monitor construction and operations of the system and, if necessary, to enforce compliance with performance standards. They may also have to support the utility itself in enforcement of industrial pretreatment regulations. Where national water pollution control standards, agencies and procedures are not fully developed, a program of institution-building should be coordinated with preparation of wastewater projects.

Monitoring

31. Because these are environmental projects, good construction inspection practices to ensure that the system is built to specifications are also good environmental management practices. Particular attention should be given to adherence to the mitigation plan provisions to protect stream channels, beaches, and wetlands. An operational monitoring program should be developed to observe trends in influent volume and strength; detect hazardous substances entering treatment works; enforce industrial pretreatment regulations; control the treatment process; assess and manage treatment plant performance; monitor environmental quality at disposal locations; and ensure sludge products and reclaimed wastewater meet reuse standards.

32. The frequency and level of sophistication of sampling depends in part on the size of the system and the nature of its treatment processes. Monitoring is expensive; it requires laboratory facilities, equipment, and technicians. As a general principle, measure only parameters necessary for managing the system, safeguarding its staff and equipment, and protecting the environment.

33. In designing the monitoring program, the emphasis should be on supporting sound operations of the wastewater system. This entails establishment of system performance standards. Data should be collected to monitor attainment of those standards, interpreted and then delivered efficiently and timely to those who must make operational decisions -- the system's operators and managers. Monitoring data is also useful to designers for improvement of future projects. All too frequently, monitoring programs are seen only, or primarily, as enforcement tools. Although enforcement action may be necessary to achieve compliance with standards in some cases, a timely report placed in the hands of a conscientious treatment plant superintendent may be more effective in protecting the environment.

Table 9.8. Wastewater Collection, Treatment, Reuse, and Disposal Systems

Potential Negative Impacts	Mitigating Measures
Direct	
1. Disturbance of stream channels, aquatic plant and animal habitat, and spawning and nursery areas during construction.	1. • Do not route sewer lines in stream channels. • Require erosion/sedimentation controls during construction.
2. Alterations in watershed hydrologic balance when wastewater is exported by collection in large upstream areas and discharge downstream.	2. • Consider sub-regional and small community systems in water-short areas. • Take full advantage of opportunities for wastewater reclamation/reuse, especially in water-short areas.
3. Degradation of neighborhoods or receiving water quality from sewer overflows, treatment works bypasses, or treatment process failure.	3. • Phase construction of collector systems and treatment works to avoid raw wastewater discharges. • Select appropriate technology. • Design for reliability, ease of maintenance. • Implement management and training recommendations, monitoring program, and industrial waste pretreatment program (see text for guidelines).
4. Degradation of receiving water quality despite normal system operation.	4. • Site and design treatment works and disposal or reuse systems on the basis of adequate data on the characteristics of the wastewater and the assimilative capacity of the receiving water body.

Table 9.8. Wastewater Collection, Treatment, Reuse, and Disposal Systems (continued)

Potential Negative Impacts	Mitigating Measures
Direct (continued)	
	• Use mathematical models for siting surface water discharges and determining required level of treatment, and for siting and designing ocean outfalls.
	• Take full advantage of appropriate land application alternatives, especially in water-short areas.
	• Implement monitoring program and industrial waste pre-treatment program (see text for guidelines).
5. Public health hazards in vicinity of discharges or reuse sites during normal operation of system.	5. • Select appropriate technology.
	• Ensure preapplication treatment and operating guidelines for land application and other water reuse systems are adequate to safeguard health of humans and livestock.
	• Restrict access to wastewater or sludge disposal sites where health hazards are unavoidable.
6. • Contamination at land application sites: • soil and crops by toxic substances and pathogens • groundwater by toxic substances and nitrogen	6. • Site and design treatment works and disposal or reuse systems on the basis of adequate data on the characteristics of the wastewater and land application site.
	• Implement monitoring program and effective industrial waste pretreatment program (see text for guidelines).
	• Ensure preapplication treatment and operating guidelines for land application and other wastewater reuse systems are adequate.

Table 9.8. Wastewater Collection, Treatment, Reuse, and Disposal Systems (continued)

Potential Negative Impacts	Mitigating Measures
Direct (continued)	
7. Failure to achieve desired beneficial uses of receiving waters despite normal system operation.	7. • Establish realistic use objective and select water quality criteria consistent with desired uses. • Establish system performance standards by modeling or other means which will result in meeting criteria.
8. Odors and noise from treatment process or sludge disposal operations.	8. • Site treatment works only near compatible land uses. • Select appropriate technology. • Include odor control and low-noise equipment in design. • Implement management and training recommendations (see text).
9. Emissions of volatile organic compounds from treatment process.	9. Establish effective industrial waste pretreatment program (see text for guidelines).
10. Soil, crop or groundwater contamination and disease vector breeding or feeding at sludge storage, reuse or disposal sites.	10. • Incorporate sludge management in system feasibility studies, technology selection, design, staffing, training, budgeting and start-up plan. • Implement effective industrial waste pretreatment program (see text for guidelines). • Ensure preapplication treatment and operating guidelines for land application and other reuse or disposal systems are adequate to safeguard health of humans and livestock. • Inspect for compliance with operating guidelines.

Table 9.8. Wastewater Collection, Treatment, Reuse, and Disposal Systems (continued)

Potential Negative Impacts	Mitigating Measures
Direct (continued)	
11. Worker accidents during construction and operation, especially in deep trenching operations.	11. Enforce adherence to safety procedures.
12. Worker accidents caused by gas accumulation in sewers and other confined spaces or by hazardous materials discharged into sewers.	12. • Emphasize safety education and training for system staff. • Implement effective industrial waste pretreatment program (see text for guidelines). • Provide appropriate safety equipment and monitoring instruments. • Enforce adherence to safety procedures.
13. Serious public and worker health hazard from chlorine accidents.	13. • Incorporate safety provisions in design, operating procedures, and training. • Prepare contingency plan for accident response.
14. Nuisances and public health hazard from sewer overflows and backups.	14. • Routinely inspect sewers for illegal connections and obstructions. • Clean sewers as necessary. • Provide monitoring system with alarms for pump station failure. • Provide alternate power supply at critical pump stations. • Educate public to prevent disposal of solid waste in sewers.

Table 9.8. Wastewater Collection, Treatment, Reuse, and Disposal Systems (continued)

Potential Negative Impacts	Mitigating Measures
Direct (continued)	
15. Failure to achieve public health improvement in serviced area.	15. Conduct sanitation and hygiene education program.
16. Dislocation of residents by plant siting.	16. Assist with resettlement (see "Involuntary Resettlement" section).
17. Perceived or actual nuisances and adverse aesthetic impacts in neighborhood of treatment works.	17. Incorporate neighborhood improvements and useful public facilities in project.
18. Accidental destruction of archaeological sites during excavation.	18. Include notification and protection procedures for cultural properties in construction contract documents (see "Cultural Property" section).
Indirect	
19. Unplanned development induced or facilitated by infrastructure.	19. • Coordinate installation of sewerage with land use planning. • Strengthen land use control regulations and institutions. • Integrate planning for infrastructure in urban development projects.
20. Regional solid waste management problems exacerbated by sludge.	20. • Incorporate sludge, excreta and septage in regional solid waste management planning and in wastewater system feasibility studies and technology selection. • Implement industrial waste pretreatment program.

Table 9.8. Wastewater Collection, Treatment, Reuse, and Disposal Systems (continued)

Potential Negative Impacts	Mitigating Measures
Indirect (continued)	
21. Loss of fisheries productivity.	21. • Evaluate importance of receiving water in local and regional fisheries. • Implement mitigating measures for direct impacts 3, 4 and 7.
22. Reduction of tourist or recreational activity.	22. • Give special attention to real or perceived nuisances and aesthetic impacts in selecting site and technology. • Implement mitigating measures for direct impacts 3, 4, 5, 7, 8 and 14.

References

Public Health and Safety

American Public Health Association. 1980. Guidelines for Managers: Evaluation of Primary Health Care in Developing Countries. Washington, D.C.: International Health Programs.

Beneson, A. S., ed. 1981. Control of Communicable Diseases in Man. 13th edition. Washington, D.C.: American Public Health Association.

Clinton, J. 1979. Health Population and Nutrition Systems in LDC's: A Manual for Rural Health Workers. Washington, D.C.: Family Health Care, Inc.

Griffin, C. C. 1989. Strengthening Health Services in Developing Countries through the Private Sector. IFC Discussion Paper 4. Washington, D.C.: World Bank.

Howe, G. M., F. King, and S. Martodipoero. 1978. Primary Child Care: A Manual for Health Workers. New Delhi, India: Oxford University Press.

Kreimer, A. and Zador, M., eds. 1989. "Colloquium on Disasters, Sustainability and Development: A Look to the 1990's." Environment Working Paper 23, Policy and Research Division. World Bank, Washington, D.C.

Manson-Bahr, P. E. C. 1982. Manson's Tropical Diseases. 18th edition. London, United Kingdom: Bailliere Tindall.

McGlashan, N. D. 1983. Geographical Aspects of Health. London, United Kingdom: Academic Press.

McJunkin, E. F. 1982. Water and Human Health. Washington, D.C.: United States Agency for International Development, Information Center.

Ukoli, F. M. A. 1984. Introduction to Parasitology in Tropical Africa. Chichester, United Kingdom: John Wiley and Sons.

Environmental Considerations for Development Projects in Urban Areas

Dunkerley, H. B., ed. 1983. Urban Land Policy: Issues and Opportunities. New York: Oxford University Press.

Gupta, D. B. 1985. "Urban Housing in India." World Bank Staff Working Paper 730. World Bank, Washington, D.C.

Hamrin, Robert D. 1990. "Policy Control Options for Comparative Air Pollution Study in Urban Areas." Environment Department Working Paper 28, Policy and Research Division. World Bank, Washington, D.C.

Krupnick, Alan and Sebastian, I. 1990. "Issues in Urban Air Pollution: Review of the Beijing Case." Environment Working Paper 1990-11, Policy and Research Division. World Bank, Washington, D.C.

Mohan, R. 1979. Urban Economic and Planning Models: Assessing the Potential for Cities in Developing Countries. Baltimore, Maryland: The Johns Hopkins University Press.

Scott, I. 1982. Urban and Spatial Development in Mexico. Baltimore, Maryland: The Johns Hopkins University Press.

Roads and Highways

Cook, C., H. L. Beenhakker, and R. E. Hartwig. 1985. "Institutional Considerations in Rural Roads Projects." World Bank Staff Working Paper 748. World Bank, Washington, D.C.

Parizek, R. R. 1971. "Impact of Highways on the Hydrogeologic Environment." In Environmental Geomorphology, edited by R. Coates. Binghamton: State University of New York.

Patterson, W. D. O. 1987. Road Deterioration and Maintenance Effects: Models for Planning and Management. Washington, D.C.: World Bank.

Watanatada, T. and others. 1987. Highway Design and Maintenance Standards Model. 2 Volumes. Washington, D.C.: World Bank.

World Bank. 1982. "Highway Sector Lending." Operational Policy Note, 10.03. World Bank, Washington, D.C.

____. 1988. Road Deterioration in Developing Countries: Causes and Remedies. World Bank Policy Study. Washington, D.C.: World Bank.

Inland Navigation

Inland Resources Foundation. 1986. Environmental Assessment Report on the Proposed Southeast Peninsula Access Road: St. Kitts, West Indies. St. Thomas, Virgin Islands.

MacKnight, S. and others. 1989. The Environmentally Sound Disposal of Dredged Materials. INU Report 54, Policy Planning and Research Staff, Infrastructure and Urban Development Department. Washington, D.C.: World Bank.

National Research Council. 1985. Dredging Coastal Ports: An Assessment of the Issues. Washington, D.C.: National Academy Press.

United States Army Corps of Engineers. 1983. Environmental Engineering for Deep-Draft Navigation Projects. Engineer Manual EM 110-2-1202. Cincinnati, Ohio: Center for Environmental Research Information.

Port and Harbor Facilities

Frankel, E. G. 1987. Port Planning and Development. New York: John Wiley and Sons.

International Association of Ports and Harbors. 1989. Guidelines for Environmental Planning and Management in Ports and Coastal Area Developments. Tokyo, Japan.

National Environment Board. 1979. Manual of Guidelines for Preparation of Environmental Impact Evaluations. Bangkok, Thailand.

Rau, J. G., and D. C. Wooten. 1980. Environmental Impact Analysis Handbook. New York: McGraw-Hill.

United States Army Corps of Engineers. 1978. Effects of Suspended Material on Aquatic Animals. Technical Report D-78-29. Cincinnati, Ohio.

Large-Scale Housing Projects

Alexander, C. 1975. The Oregon Experiment. New York: Oxford University Press.

___. 1985. The Production of Houses. New York: Oxford University Press.

Fathy, H. 1986. Natural Energy and Vernacular Architecture. Illinois: University of Chicago Press.

Goodland, R. 1977. Buildings and the Environment. Millbrook, New York: Cary Ecosystem Center.

Grimes, O. F., Jr. 1976. Housing for Low-Income Urban Families: Economics and Policy in the Developing World. Baltimore, Maryland: The Johns Hopkins University Press.

Malpezzi, S., S. K. Mayo, and D. J. Gross. "Housing Demand in Developing Countries." World Bank Staff Working Paper 733. World Bank, Washington, D.C.

Solid Waste Collection and Disposal Systems

Cointreau, S. 1982. Environmental Management of Urban Solid Wastes in Developing Countries: A Project Guide. Urban Technical Series Paper 5. Washington, D.C.: World Bank.

Cointreau, S. and others. 1985. Recycling of Municipal Refuse: State of the Art Review. World Bank Technical Paper 30. Washington, D.C.: World Bank.

Diaz, L. R., G. M., Savage, and C. G. Golueke. 1982. Resource Recovery from Municipal Solid Wastes. Boca Raton, Florida: CRC Press.

Gunnerson, C. G., ed. 1988. Wastewater Management for Coastal Cities: The Ocean Disposal Option. World Bank Technical Paper 77. Washington, D.C.: World Bank.

Holmes, J. R., ed. 1984. Managing Wastes in Developing Countries. New York: John Wiley and Sons.

Lund, R. T. 1985. Remanufacturing: The Experience of the United States and Implications for Developing Countries. World Bank Technical Paper 31. Washington, D.C.: World Bank.

Northeim, C. M. and others. 1987. Handbook for the Design, Construction, and Operation of Sanitary Landfills. Cincinnati, Ohio: United States Environmental Protection Agency, Hazardous Waste and Environmental Risk Laboratory.

Wilson, D. C. 1981. Harwell Laboratory, Waste Management: Planning, Evaluation, Technologies. Oxford, United Kingdom: Clarendon Press.

Tourism Development

Beekhuis, J. V. 1981. "Tourism in the Caribbean: Impacts on the Economic, Social, and Natural Environments." Ambio 10(6):325-331.

Cambers, G. 1985. "A Major Tourist Development on the West Coast of Barbados at Heywoods." Proceedings of the Caribbean Seminar on Environmental Impact Assessment. Barbados, West Indies.

Douglas, J. E. 1988. "The Role of the Social Contract in Effective Development Planning: The Case of the Negril Region, Jamaica." Proceedings of the International Workshop on Impact Assessment for International Development, May 31-June 4, 1987. Vancouver, British Columbia.

Pearce, D. G., and R. M. Kirk. 1986. "Carrying Capacities for Coastal Tourism." Industry and Environment 9(1):3-6.

Singh, T. V., and J. Kaur. 1986. "The Paradox of Mountain Tourism: Case References from the Himalaya." Industry and Environment 9(1): 21-26.

Tangi, M. 1977. "Tourism and the Environment." Ambio 6(6)336-341.

Western, D. 1986. "Tourism Capacity in East African Parks." Industry and Environment 9(1):14-16.

Wastewater Collection, Treatment, Reuse, and Disposal Systems

Fair, G. M., J. C. Geyer, and D. A. Okun. 1966. Water and Wastewater Engineering. 2 Volumes. New York: John Wiley and Sons.

Feachem, R. G. and others. 1983. Sanitation and Disease: Health Effects of Excreta and Wastewater Management. Chichester, United Kingdom: John Wiley and Sons.

Feachem, R. G., D. D. Mara, and M. G. McGarry. 1977. Water, Wastes and Health in Hot Climates. New York: John Wiley and Sons.

Grover, B., N. Burnett, and M. McGarry. 1983. Water Supply and Sanitation Project Preparation Handbook. 3 Volumes. Washington, D.C.

Kalbermatten, J. D., D. A. S. Julius, and C. G. Gunnerson. 1980. Appropriate Technology for Water Supply and Sanitation: A Summary of Technical and Economic Options. Washington, D.C.: World Bank.

McJunkin, E. F. 1982. Water and Human Health. Prepared by the National Demonstration Water Project for the United States Agency for International Development. Washington, D.C.: Development Information Center.

Palange, R. C., and A. Zavala. 1987. Water Pollution Control: Guidelines for Project Planning and Financing. World Bank Technical Paper 73. Washington, D.C.: World Bank.

Pettygrove, G. S., and T. Asano, eds. 1985. Irrigation with Reclaimed Municipal Wastewater--A Guidance Manual. Chelsea, United Kingdom: Lewis Publishers, Inc.

World Health Organization. 1989. Health Guidelines for Use of Wastewater in Agriculture and Aquaculture. Technical Paper Series 778. Geneva, Switzerland.

ANNEX 9-1

Sample Terms of Reference (TOR)
An Environmental Assessment of Inland Navigation

Note: Paragraph number corresponds to those in the Sample Terms of Reference (TOR) Outline in Annex 1-3; additional paragraphs are not numbered

7. Task 1. Description of the Proposed Project. Project-specific descriptions of the following:

- Disposal options for dredged materials should be specified as open-water, shoreline or upland, and whether treatment or containment will be necessary, especially if the materials are contaminated.

- Dredging equipment requirements based on physical and chemical character of dredge material, disposal site, and physical conditions at the dredge site.

- Transportation requirements such as terminal facilities and berths, barges and vessels, pipelines, roads, and disruption of transit for commercial and recreational users.

- Energy-producing operations, including solid wastes, emissions, and discharges and their quantities, and transport and handling procedures for fuels and other hazardous materials.

9. Task 3. Legislative and Regulatory Considerations. If open-water disposal in international waters is being considered then the requirements of the London Dumping Convention of 1972, Oslo Convention of 1974, and Paris Convention of 1978 should be evaluated. In addition, review of the Great Lakes Water Quality Agreement of 1978 between the United States and Canada should be consulted for scoping dredging projects.

10. Task 4. Determination of the Potential Impacts of the Proposed Project.

- Dredge Material Characterization Studies. Studies of the physical and chemical character of the dredge materials is important for selecting appropriate dredging equipment and disposal options,estimating quantities of different materials, and designing monitoring programs and mitigative measures. All sampling results should be statistically representative of the project area (i.e., vertical and horizontal extent) to ensure proper environmental planning and reduce the likelihood of encountering contaminated "hot spots".

- Screening for Sensitive Environments. Field evaluations of the ecosystems and communities in the project area environs will provide information concerning direct and indirect impacts on pervasive, sensitive, and/or threatened and endangered components of potentially affected systems. For example, alteration of flow within an aquatic system could yield adverse impacts on a downstream mangrove.

17. Consulting Team. The consultant team for preparation of an EIA for inland navigation projects should include members from the following professional disciplines: environmental impact specialist, team leader; civil/environmental engineer with experience in dredging projects and water-based transport; aquatic or marine biologist, or other specialty depending on the dredging site and natural resources which could be impacted; surface water hydrologist; land-use planner; and socioeconomist.

ANNEX 9-2

Sample Terms of Reference (TOR)
An Environmental Assessment of Port and Harbor Facilities

Note: Paragraph number corresponds to those in the Sample Terms of Reference (TOR) Outline in Annex 1-3; additional paragraphs are not numbered

7. Task 1. Description of the Proposed Project. Project-specific descriptions of the following:

- Disposal options for dredged materials should be specified as open-water, shoreline or upland, and whether treatment or containment will be necessary, especially if the materials are contaminated.
- Dredging equipment requirements based on physical and chemical character of dredge material, disposal site, and physical conditions at the dredge site.
- Transportation requirements such as terminal facilities and berths, barges and vessels, pipelines, roads, and disruption of transit for commercial and recreational users.
- Energy-producing operations, including solid wastes, emissions, and discharges and their quantities, and transport and handling procedures for fuels and other hazardous materials.
- Port-related industries which may be developed and serviced by the facilities including hazardous materials, handling, storage, processes, and disposal, special energy needs, and waste disposal requirements.

9. Task 3. Legislative and Regulatory Considerations. If open-water disposal in international waters is being considered then the requirements of the London Dumping Convention of 1972, Oslo Convention of 1974, and Paris Convention of 1978 should be evaluated. In addition, the International Maritime Organization (IMO) should be consulted for guidelines pertaining to releases and discharges from ships.

10. Task 4. Determination of the Potential Impacts of the Proposed Project.

- Dredge Material Characterization Studies. Studies of the physical and chemical character of the dredge materials is important for selecting appropriate dredging equipment and disposal options, estimating quantities of different materials, and designing monitoring programs and mitigative measures. All sampling results should be statistically representative of the project area (i.e., vertical and horizontal extent) to ensure proper environmental planning and reduce the likelihood of encountering contaminated "hot spots".

- Screening for Sensitive Environments. Field evaluations of the ecosystems and communities in the project area environs will provide information concerning direct and indirect impacts on pervasive, sensitive, and/or threatened and endangered components of potentially affected systems. For example, alteration of flow within an aquatic system could yield adverse impacts on a downstream mangrove.

- Effluent Characterization Studies. These may be necessary to design appropriate water pollution control options for the port facilities and industries. Ideally, one should identify sanitary wastewater streams, cooling water streams, and process water streams.

- "Null Zone" Identification. The existing location and potential relocation from development activities should be determined to mitigate the rate of sedimentation and saltwater intrusion as well as reduce the frequency of maintenance dredging activity.

17. Consulting Team. The consultant team for preparation of an EIA for inland navigation projects should include members from the following professional disciplines: environmental impact specialist, team leader; civil/environmental/port engineer(s) with experience in dredging projects, port and harbor development, and water-based transport; aquatic or marine biologist, or other speciality depending on the dredging site and natural resources which could be impacted; surface water hydrologist; land-use planner; and socioeconomist.

ANNEX 9-3

Sample Terms of Reference (TOR)
An Environmental Assessment of Solid Waste Disposal Systems

Note: Paragraph number corresponds to those in the Sample Terms of Reference (TOR) Outline in Annex 1-3; additional paragraphs are not numbered

7. Task 1. Description of the Proposed Project.

(a) For project improvements to solid waste collectionl, include: physical layout of the neighborhoods to receive improved collection; social, cultural and economic conditions of the neighborhoods to receive improved collection; and description of the project elements, including method of collection proposed, pilot tests to confirm the proposed collection method as appropriate, pre-implementation activities of public education and involvement, cost recovery systems, equipment specifications and procurement plans, implementation plans, operation and maintenance procedures, responsible parties for each aspect of the system.

(b) For project improvements to solid waste transfer and disposal, include: physical layout of the overall urban area to be served by transfer and/or disposal facilities, including mapping of all major roads; strategic siting of the facilities, including economic justification for the overall strategic plan of collection service areas, direct haul routes, transfer stations, transfer routes and disposal locations; physical, ecological and demographic setting of facilities, including surrounding land use characteristics, proximity to residential neighborhoods, location of public water supply sources and private wells, direction of ground water flow, uses of surface waters, prevailing wind direction; and description of the project elements, including layout of proposed facilities (e.g., fencing, buildings, weighbridges, roads, ramps, drainage, gas and leachate control systems, monitoring wells); construction schedule, operating plans, closure plans, long-term monitoring plans, and responsible parties.

8. Task 2. Description of the Environment.

(a) For project improvements to collection systems:

Physical environment: neighborhood layout, showing locations for communal containers, stops for truck during block collection, or streets served by curb-side collection; conditions of road or walkway access for collection equipment; and climate and meteorology, as it affects refuse containment and frequency of collection.

Socio-cultural environment: population density and demographic level by neighborhood; community structure of local leaders and traditional public involvement process; employment and other activities indicating patterns of movement to and from neighborhood; education level with regard to sanitation and public health; and customs and attitudes relative to cooperation with collection system.

(b) For project improvements to transfer and disposal facilities:

Physical environment: location of proposed facilities with regard to nature of surrounding land uses and proximity to homes and other establishments; existing road and traffic conditions in the area of proposed facilities, versus proposed road and traffic conditions; existing topography and proposed changes, including area which will be affected by any visible aesthetic impacts; soils and geology; surface and ground water hydrology, and hydraulic connections between the proposed sites and receiving waters downgradient of the sites; existing and proposed uses of receiving waters, including location of private and public water supply wells and intakes; climate and meteorology, including prevailing wind direction.

Biological environment: flora and fauna; sensitive habitats (e.g., wetlands delineation); and rare, endangered, or commercially important species.

Socio-cultural environment: past uses of sites and consideration of any historic significance; land use and demographic character of surrounding neighborhoods; planned development activities; education, awareness, and sensitivity of public to proposed siting of facilities; and public concerns over traffic, insects, noise, dust, odor, smoke, or aesthetic issues.

9. Task 3. Legislative and Regulatory Considerations:

- Describe national laws and local ordinances which delineate the solid waste management responsibility and authority delegated to local government. Describe national laws and guidelines which define the design and operating standards which local governments are to meet in the conduct of their responsibilities. Include description of any environmental standards which are to be met, including any requirements for submission of environmental monitoring data or environmental impact assessment statements by local governments to the national government. Describe local ordinances which govern citizen responsibility to participate in and cooperate with the solid waste system.

- Discuss the extent to which the local government uses education, inspection and enforcement to assure compliance with the available regulations. Describe the technical assistance, environmental monitoring, and regulatory enforcement activities provided by national and provisional government as a support to local government operations and actions.

10. Task 4. Determination of the Potential Impacts of the Proposed Project.

- For solid waste projects, there are numerous potential impacts to be reviewed as a part of design. For the most part, well conceived designs will minimize adverse impacts. Also, many potential impacts can be minimized by altering operating practices.

- There are some potential impact issues whose consequences would be environmentally significant over the long term. With regard to these impact issues, special studies conducted as a part of environmental impact assessment are recommended. Specifically, prior to design of a land disposal site, borings need to be drilled both on-site and off-site to assess the character of soils and geology and confirm the flow of ground water. Data from these borings coupled with information on rainfall and infiltration should be used to make a simple determination of the quantity of leachate which could be generated and released from the land disposal site and its potential effect on the nearest receiving water.

14. Task 8. Development of a Monitoring Plan.

- For solid waste projects which include a land disposal facility, environmental monitoring should include gas and ground water monitoring wells and a regular schedule of monitoring for key indicators of contamination. If the land disposal site has a gas collection and ventilation system, periodic monitoring of the composition of gas being discharged from the vents is recommended. Also recommended is periodic monitoring, on-site and off-site with a portable meter, of the ambient air's oxygen and combustible gas levels. Similarly, for projects which include an incinerator or resource recovery plan, environmental monitoring should include air quality monitoring of stack gases.

17. Consulting Team. For solid waste projects, an optimum consulting team would include the following, in order of priority: civil engineers with experience in solid waste collection and disposal; hydrogeologists with experience in ground water pollution control; sociologists or psychologists with experience in community participation in project design and operation; land use planners with experience in facility siting; biologists with environmental assessment experience in facility siting; and meteorologists with experience in air pollution control.

ANNEX 9-4

Sample Terms of Reference (TOR)
An Environmental Assessment of Tourism Development

Note: Paragraph number corresponds to those in the Sample Terms of Reference (TOR) Outline in Annex 1-3; additional paragraphs are not numbered

7. <u>Task 1. Description of the Proposed Project</u>. Provide a full description of the project and its existing setting, using maps at appropriate scales.

The proposed project should include: general layout (size, capacity, etc.); preconstruction and construction activities; operation and maintenance; life span; plans for providing utility, waste disposal, and other necessary services; physical setting, ecological setting, demographic setting, sociocultural setting and institutional setting.

8. <u>Task 2. Description of the Environment</u>. If the tourism development is associated with an existing or planned park or reserve, include copies of park/reserve management plans, appropriate maps, and special studies characterizing the resources at issue.

9. <u>Task 3. Legislative and Regulatory Considerations</u>. Describe the pertinent regulations and standards governing environmental quality, health and safety, protection of sensitive areas, protection of endangered species, siting, land use control, rights of indigenous peoples, etc., at international, national, regional and local levels.

10. <u>Task 4. Determination of the Potential Impacts of the Proposed Project</u>. Special studies may include the following:

- Environmental carrying capacity of sensitive ecological sites or cultured properties.
- Social carrying capacity, including attitudes of local people to the proposed influx of foreigners and potential sources of conflict.
- Physical carrying capacity of local infrastructure and public services (if not adequately addressed in feasibility studies).

12. Task 9. Assist in Inter-Agency Coordination and Public/NGO Participation. In tourism projects, it is critical to involve all potentially involved government agencies, especially at the local level. It is also important to provide complete information to the affected community, so that community members can form their opinions about the project.

17. Consulting Team. A typical EA team may have any or all of the following disciplines: environmental impact specialist, team leader; civil engineer: wastewater, roads, ports and harbors, water supply; ecologist; cultural specialist such as park and recreation planner; urban sociologist or anthropologist; specialist in tourism supply/demand analysis; and legal expert(s) on land tenure, environmental law, cultural property protection.

ANNEX 9-5

Sample Terms of Reference (TOR)
An Environmental Assessment of Wastewater Collection, Treatment, Reuse, and Disposal Systems

Note: Paragraph number corresponds to those in the Sample Terms of Reference (TOR) Outline in Annex 1-3; additional paragraphs are not numbered

5. Study Area. The study area for the assessment consists of the drainage area to be serviced by the wastewater collection system; the tracts of land on which effluent or sludge are to be applied in reuse systems; marine, estuarine or inland waters which could be influenced by effluent discharge; remote sites identified for disposal of solid waste generated in the treatment process; and, if incineration is included as a sludge disposal technique, the airshed which might be affected.

7. Task 1. Description of the Proposed Project. Provide a full description of the project: location; general layout; unit process description and diagram; size in terms of population and population equivalents, present and projected; number and types of connected industries; anticipated influent and effluent characteristics; preconstruction and construction activities; schedule, staffing and support facilities and services; operation and maintenance activities; required off-site investments; and life span.

8. Task 2. Description of the Environment. Assemble, evaluate and present baseline data on the environmental characteristics of the study area. Include information on any changes anticipated before the project commences.

(a) Physical environment: geology (general description for overall study area and details for land application sites); topography; soils (general description for overall study area and details for land application sites); monthly average temperatures, rainfall and runoff characteristics; description of receiving waters (identity of streams, lakes, or marine waters; annual average discharge or current data by month, chemical quality; existing discharges or withdrawals).

(b) Biological environment: terrestrial communities in areas affected by construction, facility siting, land application or disposal; aquatic, estuarine or marine communities in affected waters; rare or endangered species; sensitive habitats, including parks or preserves, significant natural sites; species of commercial importance in land application sites and receiving waters.

(c) Sociocultural environment: present and projected population; present land use; planned development activities; community structure; present and projected employment by industrial category; distribution of income, goods and services; recreation; public health; cultural properties; indigenous peoples; customs, aspirations and attitudes.

9. Task 3. Legislative and Regulatory Considerations. Describe the pertinent regulations and standards governing environmental quality, pollutant discharges to surface waters and land, industrial discharges to public sewers, water reclamation and reuse, agricultural and landscape use of sludge, health and safety, protection of sensitive areas, protection of endangered species, siting, land use control, etc., at international, national, regional and local levels (The TOR should specify those that are known and require the consultant to investigate for others.)

10. Task 4. Determination of the Potential Impacts of the Proposed Project. In this analysis, distinguish between significant positive and negative impacts, direct and indirect impacts, and immediate and long-term impacts. Identify impacts which are unavoidable or irreversible. Wherever possible, describe impacts quantitatively, in terms of environmental costs and benefits. Assign economic values when feasible. Characterize the extent and quality of available data, explaining significant information deficiencies and any uncertainties associated with predictions of impact. If possible, give the TOR for studies to obtain the missing information.

Special attention should be given to:

- The extent to which receiving water quality standards and/or beneficial use objectives will be achieved with the proposed type and level of treatment.
- The length of stream or expanse of lake or marine waters which will be positively or negatively affected by the discharge, and the magnitude of the changes in water quality parameters.
- Projected quantitative changes in beneficial uses, such as fisheries (species composition, productivity), recreation and tourism (visitor-days, overnights, expenditures), and waters available for portable supply, irrigation, and industrial use.
- Sanitation and public health benefits anticipated.

11. Task 5. Analysis of Alternatives to the Proposed Project. Describe alternatives that were examined in the course of developing the proposed project and identify other alternatives which would achieve the same objectives. The concept of alternatives extends to siting and design, technology selection, construction techniques and phasing, and operating and maintenance procedures. Compare alternatives in terms of potential environmental impacts, land and energy requirements, capital and operating costs, reliability, suitability under local conditions, and in-

stitutional, training, and monitoring requirements. When describing the impacts, indicate which are irreversible or unavoidable and which can be mitigated. To the extent possible, quantify the costs and benefits of each alternative, incorporating the estimated costs of any associated mitigating measures. Include the alternative of not constructing the project, in order to demonstrate environmental conditions without it.

12. Task 6. Development of Management Plan to Mitigate Negative Impacts. Recommend feasible and cost-effective measures to prevent or reduce significant negative impacts to acceptable levels. Estimate the impacts and costs of those measures, and of the institutional and training requirements to implement them. Consider compensation to affected parties for impacts which cannot be mitigated. Prepare a management plan including proposed work programs, budget estimates, schedules, staffing and training requirements, and other necessary support services to implement the mitigating measures.

17. Consulting Team. The following specialties should be represented on the core consulting team: environmental engineering, environmental planning (or other environmental generalists); ecology (terrestrial, aquatic or marine, depending on type of discharge); water quality; soils science (for land application); wastewater utility management; and sociology/anthropology. Other specialties that may be needed depending on the nature of the project are public health, agronomy, hydrology, land use planning, oceanography, water quality modelling, and resource economics.

Specify dates for progress reviews, interim and final reports, and other significant events.

19. Other Information. Include here lists of data sources, project background reports and studies, relevant publications, and other items to which the consultant's attention should be directed. Examples are pre-feasibility studies, population and land use projections, land use plans, industrial activity information, water quality studies, sewerage service needs surveys, public health reports, sewer system evaluations.

ABBREVIATIONS/ACRONYMS

ACC/SCN	(U.N.) Administrative Committee on Co-ordination/Sub Committee on Nutrition
ADB	African Development Bank (see AfDB) and Asian Development Bank (see AsDB)
AfDB	African Development Bank (see ADB)
AGR	Sociological Advisor
AID	(United States) Agency for International Development (see USAID)
API	American Petroleum Institute
AsDB	Asian Development Bank (see ADB)
ASEAN	Association of Southeast Asian Nations
ATL	Action Threshold Level
BOD	Biochemical Oxygen Demand
BOD_5	Biochemical Oxygen Demand Over Five Days
BTO	Back-To-Office Report
C	Carbon
°C	(Centigrade) Celsius
CBA	Cost-Benefit Analysis
CD	Country Department
CECC	Credit Extension Coordinating Committee
CFC	Chlorofluorocarbons
CFR	Code of Federal Regulations
CGIAR	Consultative Group on International Agricultural Research
CH_4	Methane
CIDA	Canadian International Development Agency
CITES	(UN) Convention on International Trade in Wild Flora and Fauna
CO	Carbon monoxide
CO_2	Carbon Dioxide
COD	Chemical Oxygen Demand Country Operations Division
CMEA	Council for Mutual Economic Assistance
CSI	Construction Specifications Institute
CSP	Country Strategy Paper
DANIDA	Danish International Development Agency
DAP	diammonium phosphate
dB	decibel
DCCI	Development Commission for Cement Industry (India)
DFI	Development Finance Institution
DMG	Drylands Management Guidelines
DNA	Deoxyribonucleic Acid
EA	Environmental Assessment
EA OD	Environmental Assessment Operational Directive

EAPs	Environmental Action Plans
EAR	Environmental Assessment Report
EBRD	European Bank for Reconstruction and Development
EC	European Communities
ECU	European currency unit
EDF	European Development Fund
EDI	Economic Development Institute of the World Bank
EDP	Environmentally-adjusted Net Domestic Product
EEC	European Economic Community
EIB	European Investment Bank
EIP	Environmental Issues Paper
EMF	Electro Magnetic Field
ENV	Environment Department
EPA	(United States) Environmental Protection Agency (see USEPA)
EPD	Environmental Protection Department
EPS	Executive Project Summary
ER	Environmental Review
ERL	Emergency Recovery Loan Emergency Reconstruction Loan
ERR	Economic Rate of Return
ESCAP	Economic and Social Commission for Asia and the Pacific
ESMAP	Energy Sector Management Assistance Program
EXTIE	International Economic Relations Division of the External Affairs Department (World Bank)
°F	Fahrenheit
FAO	Food and Agricultural Organization of the United Nations
FEPA	(Nigerian) Federal Environmental Protection Agency
FEPS	Final Executive Project Summary
FI	Financial Intermediary
FIL	Financial Intermediary Loan
FINNIDA	Finnish Department of International Development Cooperation
FMWH	Federal Ministry of Works and Housing
FY	Fiscal Year
GATT	(UN) General Agreement on Tariffs and Trade
GDP	Gross Domestic Product
GEF	Global Environment Facility
GLC	ground level concentrations
GNP	Gross National Product
GOI	Government of India
GOR	Government of Rwanda
GTZ	German Agency for Technical Cooperation
ha	hectare
HABITAT	United Nations Centre for Human Settlements
HHS	(United States) Department of Health and Human Services

H_2S	Hydrogen Sulfide
IADB	Inter-American Development Bank
IBRD	International Bank for Reconstruction and Development
ICICI	Industrial Credit and Investment Corporation of India Limited
ICCROM	International Centre for the Study of the Preservation and the Restoration of Cultural Property
ICOMOS	International Committee of Monuments and Sites
IDA	International Development Association
IDB	Inter-American Development Bank
IDBI	Industrial Development Bank of India
IEC	information, education, and communication
IEPS	Initial Executive Project Summary
IFAD	(UN) International Fund for Agricultural Development
IFC	International Finance Corporation of the World Bank
IIED	International Institute for Economic Development
ILO	(UN) International Labour Organization
IMF	International Monetary Fund
IMO	International Maritime Organization
IPCC	Intergovernmental Panel on Climate Change
IPM	Integrated Pest Management
IRR	Internal Rate of Return
ITTO	International Tropical Timber Organization
IUCN	International Union for Conservation of Nature and Natural Resources
km	kilometer
km^2	square kilometer
KV	kilovolts
kW	kilowatts
m	meter
MAP	monoammonium phosphate
mm	millimeter
MIGA	Multilateral Investment Guarantee Agency of the World Bank
Miniplan	Ministry of Plan
MOE	Ministry of Environmental Protection
MOP	Memorandum of the President
MOS	Monthly Operational Summary
MW	Mega Watts
NEAP	National Environmental Action Plan
NEC	National Environment Commission
NGO	Nongovernmental Organization
NIOSH	National Institute for Occupational Safety and Health
NORAD	Norwegian Agency for International Development
NO_x	Oxides of Nitrogen
NPV	Net Present Value
N_2O	Nitrous Oxide

O, O_2, O_3	Oxygen
OAU	Organization for African Unity
OD	Operational Directive
ODA	Overseas Development Administration (United Kingdom)
OECD	Organization for Economic Cooperation and Development
OED	Operations Evaluation Department
O/G	oil and grease
OIEC	Organization for International Cooperation
OMS	Operational Manual Statement
OPN	Operations Policy Note
OPNSV	Senior Vice President, Operations
PB	Project Brief
PCB	polychlorinated biphenyl
PCR	Project Completion Report
PEPA	Pakistan Environmental Protection Agency
pH	measurement of acidity and alkalinity (on log. scale 0-14, 7 = neutral, ≤ 7 = increasing acidity, ≥ 7 = increasing alkalinity)
PHN	Population, Health and Nutrition Department (World Bank)
PIDs	Provincial Irrigation Departments
PI/ER	Public Investment/Expenditure Review
PIP	Public Investment Program
PIR	Project Implementation Review
POPTR	Personnel Operations Department, Training Division
ppb	parts per billion
PPF	Project Preparation Facility
PPR	Project Performance Report
PR	President's Report
PRC	People' Republic of China
PRE	Policy, Research, and External Affairs
R&D	Research and Development
RED	Regional Environment Division
ROW	Right-of-Way
RVP	Regional Vice President
SAL	Structural Adjustment Lending
	Structural Adjustment Loan
SAR	Staff Appraisal Report
SCBA	Social Cost Benefit Analysis
SDC	Swiss Development Corporation
SECAL	Sector Adjustment Loan
SIDA	Swedish International Development Authority
SNA	System of National Accounts
SOD	Sector Operations Division
SO_2	Sulfur Dioxide
SO_x	Oxides of Sulfur

SPPF	Special Project Preparation Facility
SPRIE	International Economic Relations Division of the Strategic Planning Department
STEL	Short-Term Exposure Limit
TAL	Technical Assistance Loan
TDS	Total Dissolved Solids
TFAP	Tropical Forest Action Plan
TLV	Threshold Limit Values
TM	Task Manager
TMA	time weighted average
TOC	Total Organic Carbon
TOR	Terms of Reference
TSD	Toxic Storage Disposal
TSP	Total Suspended Particulates
TSS	Total Suspended Solids
UNDP	United Nations Development Programme
UNDRO	United Nations Disaster Relief Organization
UNEP	United Nations Environment Programme
UNESCO	United Nations Educational, Scientific and Cultural Organization
UNICEF	United Nations Children's Fund
UNIDO	United Nations Industrial Development Organization
UNSO	United Nations Statistical Office
USACE	United States Army Corps of Engineers
USAID	United States Agency for International Development (see AID)
USEPA	United States Environmental Protection Agency (see EPA)
WAPDA	Water and Power Development Authority
WHO	World Health Organization
WMA	Wildlands Management Area
WUAs	Water User Associations

ENVIRONMENTAL ASSESSMENT

A Guide To Further Reading

Abel, N., and M. Stocking. 1981. "The Environmental Assessment Experience of Underdeveloped Countries." In Project Appraisal and Policy Review, edited by T. O'Riordan and W. R. D. Sewell. Chichester, United Kingdom: John Wiley and Sons.

___. 1990. Environmental Risk Assessment: Dealing with Uncertainty in Environmental Impact Assessment. Environment Paper 7. Manila, Philippines: Office of the Environment.

Ahmad, Y.J., and G. K. Sammy. 1985. Guide to Environmental Impact Assessment in Developing Countries. London: Hodder and Stoughton (for the United Nations Environment Programme).

___. 1987. Orientaciones para la Evaluacion del Impacto Ambiental en los Paises en Desarrollo. Nairobi, Kenya: PNUMA (also in French).

American Arbitration Association. 1980. Improving EIS Scoping. Washington D.C.

American Society of Civil Engineers. 1989. Guidelines for Planning and Designing Hydroelectric Developments. Volume I: Planning, Design of Dams and Related Topics, and Environmental. New York: ASCE.

___. 1989. Guidelines for Planning and Designing Hydroelectric Developments, Volume II: Waterways. New York: ASCE.

Anderson, A., ed. 1990. Alternatives to Deforestation. New York: Columbia University Press.

Andrews, R. N. L., and others. 1977. Substantive Guidance for Environmental Inpact Assessment: An Exploratory Study. Indianapolis, Indiana: Butler University, Holcomb Research Institute and the Institute of Ecology.

Anon, 1988. "The International Development of Environmental Impact Assessment." The Environmentalist 8(2):143.

___. 1988. "Transportation Elements of Environmental Impact Assessments and Reports." Institute of Transportation Engineering Journal 58(6):69-76.

Asian Development Bank. 1986. Environmental Guidelines for Selected Infrastructure Projects. 1 Volume. Manila, Philippines: Infrastructure Department, Environment Unit.

___. 1988e. Guidelines for Integrated Regional Economic-Environmental Development Planning: A Review of Regional Environmental Development Planning Studies in Asia. Paper No. 3, Volume I: Guidelines. Manila, Philippines.

___. 1988f. Guidelines for Integrated Regional-Environmental Development Planning: A Review of Regional Environmental Development Planning Studies in Asia. Paper No. 3, Volume II: Case Studies. Manila, Philippines.

___. 1987. Handbook on the Use of Pesticides in the Asia-Pacific Region. Manila, Phillipines.

Association of Southeast Asian Nations. 1985. Report of Workshop on the Evaluation of Environmental Impact Assessment Applications in ASEAN countries. Bandung, Indonesia.

Ayanda, J. O. 1988. "Incorporating Environmental Impact Assessment in the Nigerian Planning Process: Need and Procedure." Third World Planning Review 10:51-64 (U.K.).

Barbier, E. B. 1990. "Alternative Approaches to Economic-Environmental Interactions." Ecology Economics 2:7-26.

___. 1989. Economics, Natural Resource Scarcity and Development: Conventional and Alternative Views. London: Earthscan Publications Ltd.

___. 1988. "Economic Valuation of Environmental Impacts." Project Appraisal 3:143-150.

___. 1991. "Environmental Sustainability and Cost-Benefit Analysis." Environment and Planning 22:1259-1266.

Barbier, E. B., A. Markandya, and D. W. Pearce. 1990. "Sustainable Agricultural Development and Project Appraisal." European Review of Agrarian Economics 17(2): 181-196.

Barrett, B. P. D., and R. Therivel. 1990. Environmental Policy and Impact Assessment in Japan. London, United Kingdom: Routledge (in press).

Bartlett, R. V., ed. 1989. Policy Through Impact Assessment: Institutionalized Analysis as a Policy Strategy. New York: Greenwood.

Bauchum, R. G. 1985. Needs Assessment Methodologies in the Development of Impact Statements. Monticello, Illinois: Vance Bibliographies.

Becker, D.S., and J. W. Armstrong. 1988. "Development of Regionally Standardized Protocols for Marine Environmental Studies." Marine Pollution Bulletin 19(7):310-313.

Becker, H. A., and A. L. Porter, eds. 1986. Methods and Experiences in Impact Assessment. Atlanta, Georgia: International Association for Impact Assessment.

• Bellia, V., and E. D. Bidone. 1990. Rodovias, Recursos Naturais e Meio Ambiente. Rio de Janerio: Departamento Nacional de Estradas e Rodagem.

Bendix, S., and H. R. Graham, eds. 1986. Environmental Assessment: Approaching Maturity. Ann Arbor, Michigan: Ann Arbor Science.

Bisset, R. 1980. "Methods for Environmental Impact Analysis: Recent Trends and Future Prospects." Journal of Environmental Management 11:27-43.

——. 1984. "Post Development Audits to Investigate the Accuracy of Environmental Impact Predictions." Zeitschrift fur Umweltpolitik 7:463-484.

Biswas, A. K., and Qu Geping [Chu, Ko-Ping], eds. 1987. Environmental Impact Assessment for Developing Countries. Oxford, United Kingdom: Tycooly International (for the United Nations University).

Bochniarz, Z., and A. Kassenberg. 1985. Environmental Protection by Integrated Planning. Warsaw, Poland: Economic and Social Problems of Environmental Planning and Processing.

Bojo, J., K-G Maler, and L. Unemo. 1988. Economic Analysis of Environmental Consequences of Development Projects. Stockholm, Sweden: The Economic Research Institute, Stockholm School of Economics.

Bowden, M-A, and F. Curtis. 1988. "Federal EIA in Canada: EARP as an Evolving Process." Environmental Impact Assessment Review 8(1): 97-106.

Bowonder, B., and S. S. Arvind. 1989. "Environmental Regulations and Litigation in India." Project Appraisal 4:182-196.

Bregha, F. and others. 1990. The Integration of Environmental Considerations into Government Policy. Hull, Quebec: The Rawson Academy of Aquatic Science.

Burkhardt, D. F., and W. H. Ittelson, eds. 1978. Environmental Assessment of Socioeconomic Systems. New York: Plenum.

Cable, T. T., V. Brack, and V. R. Holmes. 1989. "Simplified Method for Wetland Habitat Assessment." Environmental Management 13:207-13.

Cairns, J., and T. V. Crawford, eds. 1991. Integrated Environmental Management. Chelsea, Michigan: Lewis Publishers.

Campbell, M. J. 1990. New Technology and Rural Development: The Social Impact. London, United Kingdom: Routledge.

Canter, L. W. 1977. Environmental Impact Assessment. New York: McGraw-Hill.

——. 1985. Environmental Impact of Water Resources Projects. Chelsea, Michigan: Lewis Publishers.

___. 1982. The Status of Environmental Impact Assessment in Developing Countries. Norman: University of Oklahoma, Environment and Ground Water Institute.

Canter, L. W., and L. G. Hill. 1981. Handbook of Variables for Environmental Impact Assessment. Ann Arbor, Michigan: Ann Arbor Science.

Carley, M. J., and E. O. Derow. 1983. Social Impact Assessment: A Cross-Disciplinary Guide to the Literature. Boulder, Colorado: Westview Press.

Carley, M. J., and E. S. Bustelo. 1984. Social Impact Assessment: A Guide to the Literature. Boulder, Colorado: Westview Press.

Carpenter, R. A., and J. E. Maragos. 1989. How to Assess Environmental Impacts on Tropical Islands and Coastal Areas: A Training Manual. Honolulu, Hawaii: East-West Center, Environment and Policy Institute.

Center for Environmental Management and Planning. 1986. The EEC Environmental Assessment Directive: Towards Implementation. Scotland: Aberdeen University and the United Kingdom Department of Environment.

Child, R. D. and others. 1987. Arid and Semi-Arid Rangelands: Guides for Development. Natural Resources Expanded Information Base Project. Morrelton, Arkansas: Winrock International.

Clark, B. D. and others. 1981. A Manual for the Assessment of Major Development Proposals. London, United Kingdom: HMSO.

Clark, B. D., R. Bisset, and P. Wathern. 1980. Environmental Impact Assessment: A Bibliography with Abstracts. London, United Kingdom: Mansell Publishers.

Clark, B. D. and others, eds. 1984. Perspectives on Environmental Impact Assessment. Dordrecht, The Netherlands: Riedel.

Clark, M., and J. Herrington, eds. 1988. The Role of Environmental Assessment in the Planning Process. London, United Kingdom: Mansell Publishers.

Cohrrsen, J. J., and V. T. Covello. 1989. Risk Analysis: A Guide to Principles and Methods for Analyzing Health and Environmental Risks. Washington D.C.: Council on Environmental Quality.

Conway, G. 1986. Agroecosystem Analysis for Research and Development. Bangkok, Thailand: Winrock International Institute for Agricultural Development.

Conway, G. R., ed. 1986. The Assessment of Environmental Problems. London, United Kingdom: Imperial College, Centre for Environmental Technology.

Cook, P. L. 1983. A Review of the Recent Research on the Utility of Environmental Impact Assessment. Chania, Crete: Environmental Impact Assessment Symposium.

___. 1979. Costs of Environmental Impact Statements and the Benefits they Yield to Improvements in Projects and Opportunities for Public Involvement. Villach, Austria: Economic Commission for Europe.

Covello, V. T. and others, eds. 1985. Environmental Impact Assessment and Risk Analysis: Contributions from the Psychological and Decision Sciences. New York: Springer Publishing Company.

Daly, H. E., and J. B. Cobb. 1989. For the Common Good: Redirecting the Economy toward Community, the Environment, and a Sustainable Future. Boston, Massachusetts: Beacon Press.

Davies, G. S., and F. G. Muller. 1983. A Handbook on Environmental Assessment for Use in Developing Countries. Nairobi, Kenya: UNEP.

DeJongh, P. E. 1985. Environmental Impact Assessment: Methodologies, Prediction and Uncertainty. Utrecht, The Netherlands: IAIA Congress.

___. 1985. Technical Aspects of Training in Environmental Impact Assessment, with Emphasis on Ecological Impacts. Maastricht, The Netherlands: European Institute for Public Administration.

Derman, W., and S. Whiteford. 1985. Social Impact Analysis and Development Planning in the Third World. Boulder, Colorado: Westview Press.

Dixon, J. D. and others, eds. 1988. Economic Analysis of the Environmental Impacts of Development Projects. London/Manila: Earthscan Publications Ltd. and Asian Development Bank.

Draggan, S., J. J. Cohrsson, and R. E. Morrison, eds. 1987. Environmental Monitoring, Assessment and Management. New York: Praeger.

Duinker, P. N. 1989. "Ecological Efforts Monitoring in Environmental Impact Assessment: What Can It Accomplish?" Environmental Management 13:797-805.

Eberhardt, L. L. 1976. "Quantitative Ecology and Impact Assessment." Journal of Environmental Management 4:27-70.

Economic Commission for Europe. 1990. Post-Project Analysis for Environmental Impact Analysis. New York: United Nations.

Elkin, T. J., and P. G. R. Smith. 1988. "What is a Good Environmental Impact Statement? Reviewing Screening Reports from Canada's National Parks." Journal of Environmental Management 26(1): 71-89.

Elking-Savatsky, P. D. 1986. Differential Social Impacts of Rural Resource Development. Boulder, Colorado: Westview Press.

Environmental Resources Limited. 1990. Environmental Assessment Procedures in the U.N. System. A Study prepared at the request of the United Nations System. London, United Kingdom.

Erikson, P. A. 1979. Environmental Impact Assessment: Principles and Applications. New York: Academic Press.

Evans, J. 1982. Plantation Forestry in the Tropics. Oxford, United Kingdom: Clarendon Press.

Evers, F. W. R. 1986. "Environmental Assistance and Development Assistance: The Work of the OECD." In Methods and Experiences in Impact Assessment, edited by H. A. Becker and A. L. Porter. Atlanta, Georgia: International Association for Impact Assessment.

Finnish Department of International Development Cooperation. 1989. "Guidelines for Environmental Assessment in Development Assistance." (Draft). Finland.

Finsterbusch, K., J. Ingersoll, and L. G. Llewellyn. 1990. Methods for Social Analysis in Developing Countries. Boulder, Colorado: Westview Press.

Finsterbusch, K., and C. P. Wolfe, eds. 1977. Methodology of Social Impact Assessment. Stroudsberg Pennsylvania: Dowden, Hutchinson and Ross.

Food and Agricultural Organization. 1982. Environmental Impact Analysis and Agricultural Development. FAO Environment Paper 2. Rome, Italy.

___. 1982. "Environmental Impact of Forestry." Conservation Guide 7:1-85.

___. 1977c. Planning Forest Roads and Harvesting Systems. Forestry Paper 2. Rome, Italy.

Fortlage, C.A., 1990. Environmental Assessment: A Practical Guide. Aldershot, United Kingdom: Gower.

Frideres, J. S., and J. E. DiSanto, eds. 1986. Issues of Impact Assessment: Development of Natural Resources. Atlanta, Georgia: International Association for Impact Assessment.

Gamman, J. K., and S. T. McCreary. 1988. "Suggestions for Integrating Environmental Impact Assessment and Economic Development in the Caribbean." Environmental Impact Assessment Review 8(1):43-62.

Glenn, J. C. 1988. Livestock Production in North Africa and the Middle East. Problems and Perspectives. World Bank Discussion Paper 38. Washington, D.C.: World Bank.

Go, F.C. 1987. Environmental Impact Assessment: An Analysis of the Methodological and Substantive Issues Affecting Human Health Considerations. London, United Kingdom: Monitoring and Assessment Research Centre/WHO/UNEP.

____. 1988. Environmental Impact Assessment: Operational Cost Benefit Analysis. London, United Kingdom: King's College, Monitoring and Assessment Research Center.

Goldberg, E. D., ed. 1982. Atmospheric Chemistry. Berlin: Springer Verlag.

Golden, J. and others. 1979. Environmental Impact Data Book. Ann Arbor, Michigan: Ann Arbor Science.

Gooden, P. M., and A. I. Johnstone. 1988. "Environmental Impact Assessment: Its Potential Application to Appropriate Technology in Developing Countries." The Environmentalist 8(1): 57-66.

Goodland, R. 1989. "The Environmental Implications of Major Projects in Third World Development. In Major Projects and the Environment, edited by P. Morris. Oxford, United Kingdom: Major Projects Association.

____. ed. 1990. The Race to Save the Tropics. Washington, D.C.: Island Press.

Goodland, R., C. Watson, and G. Ledec. 1985. Environmental Management in Tropical Agriculture. Boulder, Colorado: Westview Press.

Gorse, J. E., and D. R. Steeds. 1988. Desertification in the Sahelian and Sudanian Zones of West Africa. World Bank Technical Paper 61. Washington, D.C.: World Bank.

Gough, J. D. 1989. Strategic Approach to the Use of Environmental Impact Assessment and Risk Assessment Within the Decision-Making Process. Center for Resource Management Paper 13. New Zealand: University of Canterbury and Lincoln College.

Gunnerson, C. G., and D. C. Stuckey. 1986. Anaerobic Digestion: Principles and Practices for Biogas Systems. World Bank Technical Paper 49. Washington, D.C.: World Bank.

Gunnerson, C. G. 1989. Post-Audits of Environmental Programs and Projects. New York: American Society of Civil Engineers.

Hall, A. L., and J. Midgley, eds. 1988. Development Policies: Sociological Perspectives. Manchester, United Kingdom: Manchester University Press.

Hipel, K. W. 1988. "Nonsparametric Approaches to Environmental Impact Assessment." Water Resource Bulletin 24(3):487-492.

Holling, C. S., ed. 1978. Adaptive Environmental Assessment and Management. New York: John Wiley and Sons.

Horberry, J. A. J. 1984. "Development Assistance and the Environment." Ph.D. Dissertation, Massachusetts Institute of Technology. Cambridge, Massachusetts.

___. 1987. "Environmental Impact Assessment for Development." ATAS Bulletin 4:59-60.

Horstmann, K., comp. 1985. Environmental Impact Assessment for Development, ed. K. Klennert. Feldafing, Federal Republic of Germany.

Howe, G. M., ed. 1977. A World Geography of Human Diseases. New York: Academic Press.

Hufschmidt, M. M., and R. A. Carpenter. 1980. Natural Systems Assessment and Benefit-Cost Analysis for Economic Development. Honolulu, Hawaii: East-West Center.

Hufschmidt, M. M. and others. 1983. Environment, Natural Systems and Development: An Economic Development Guide. Baltimore, Maryland: The John Hopkins University.

Hunsaker, C. T. and others. 1990. "Assessing Ecological Risk on a Regional Scale." Environmental Management 14:325-332.

Hyman, E., and B. Stiftel. 1988. Combining Facts and Values in Environmental Impact Assessment: Theories and Techniques. Boulder, Colorado: Westview Press.

Ingram, G. K. 1984. "Housing Demand in the Developing Metropolis: Estimates from Bogota and Cali, Colombia." World Bank Staff Working Paper 733. World Bank, Washington, D.C.

Institute for Solid Wastes of the American Public Works Association. 1975. Solid Waste Collection Practice. Chicago, Illinois.

Inter-American Development Bank. 1991. Application of Environmental Procedures in the Agricultural Sector. Guidelines. 2nd edition. Washington, D.C.

___. 1991. Application of Environmental Procedures in the Sanitation and Urban Development Sector. Guidelines. Washington, D.C.

___. 1990. Procedures for Classifying and Evaluating Environmental Impacts of Bank Operations. Washington, D.C.

International Commission for Environmental Assessment. 1990. Final Report of the Working Group on an International Commission for Environmental Assessment. Utrecht, The Netherlands.

Jaffe, L. S. 1973. "Carbon Monoxide in the Biosphere: Sources, Distributions and Concentrations." Journal of Geophysical Research 67(5):293-305.

Janicke, M. and others. 1989. "Economic Structure and Environmental Impacts: East-West Comparisons." Environmentalist 9:171-83.

Janikowski, R., and A. Starzewska. 1986. "Environmental Impact Assessment Project in Poland." Environmental Impact Worldletter (May-June):1-4.

Khosla, P. K., and D. K. Khurana, eds. 1987. Agroforestry for Rural Needs. New Delhi, India: Indian Society of Tree Scientists.

Khan, S. A. 1987. "Social Impact of Agricultural Development in Bangladesh: A critique of the Differentiation/Polarization Thesis." Journal of Social Studies 37:15-29.

Lal, R. and others. 1983. Land Clearing in the Tropics. Boston, Massachusetts: A. A. Balkema.

Lang, R., and A. Armour. 1980. Environmental Planning Resourcebook. Ottawa, Canada: Environment Canada, Lands Directorate.

Lavine, M. J. and others. 1978. "Bridging the Gap between Economic and Environmental Concerns in Environmental Impact Assessment." Environmental Impact Assessment Review 2. New York: Elsevier Science Publishing Company.

Lee, N. 1987. Environmental Impact Assessment: A Training Guide. Department of Town and Country Planning Paper 18. Manchester, United Kingdom: University of Manchester.

___. 1982. "The Future Development of Environmental Impact Assessment." Journal of Environmental Management 14:71-90.

___. 1984. Training for Environmental Impact Assessment. Brussels, Belgium: Economic Commission for Europe.

Lee, N., and C. M. Wood. 1985. "Training for Environmental Impact Assessment within the Economic Commission for Europe." Journal of Environmental Management 21:271-286.

Lee, N., C. M. Wood, and V. Gazidellis. 1985. Arrangements for Environmental Impact Assessment and Their Training Implications in the European Communities and North America. Department Department of Town and Country Planning Paper 13. Manchester, United Kingdom: University of Manchester.

Leistritz, F. L., and B. L. Ekstrom. 1986. Social Impact Assessment and Management: An Annotated Bibliography. New York: Garland Publishers.

Leistritz, F. L., and S. H. Murdock. 1981. The Socioeconomic Impact of Resource Development. Boulder, Colorado: Westview Press.

Leopold, L. B. and others. 1971. A Procedure for Evaluating Environmental Impact. U.S. Geological Survey Circular 645. Washington, D.C.

Lichfield, N. 1989. "Environmental Impact Assessment in Project Appraisal in Britain." Project Appraisal 3:133-141.

Lohani, B. N. 1984. Environmental Quality Management. New Delhi, India: South Asian Publishers.

Macrory, R., and M. Lafontaine. 1982. Public Enquiry and Enquete Publique. London, United Kingdom: Institute for European Environmental Policy.

Metcalf & Eddy, Inc. 1972. Wastewater Engineering: Collection, Treatment, Disposal. New York: McGraw-Hill.

McCormick, J. 1985. Acid Earth: A Global Threat of Acid Pollution. Washington, D.C.: International Institute for Environment and Development.

McCracken, J. A., J. N. Pretty, and G. Conway. 1989. An Introduction to Rapid Rural Appraisal for Agricultural Development. London, United Kingdom: International Institute for Environment and Development.

McEvoy, J., and T. M. Dietz. 1977. Handbook for Environmental Planning: The Social Consequences of Environmental Change. New York: John Wiley and Sons.

Mills, J. S., and S. Diamond. 1988. "Environmental Impact Statements." Environmental Science and Technology 22(6):618-620.

Mitsch, W. J., and S. E. Jorgensen, eds. 1989. Ecological Engineering: An Introduction to Ecotechnology. New York: John Wiley and Sons.

Munn, R. E., ed. 1979. Environmental Impact Assessment: Principles and Procedures. Published on behalf of the Scientific Committee on Problems of the Environment of the International Council of Scientific Union. Chichester, United Kingdom: John Wiley and Sons.

Murdock, S. H. and others. 1982. "An Assessment of Socio-Economic Assessments: Utility, Accuracy and Policy Considerations." Environmental Impact Assessment 3(4):333-350.

Murthy, K. S. 1988. National Environmental Policy Act Process. Boca Raton, Florida: CRC Press.

National Technical Information Service. 1990. Toxicity of Fibers and Fiber Composition: A Bibliography. Document No. PB 90-87555. Washington, D.C.

Neto, Z. F. de Lacerda. 1989. "Environmental Impact Assessment in Brazil." Environmental Education and Information 8(2):96-99.

Organization of American States. 1984. Integrated Regional Development Planning: Guidelines and Case Studies from OAS Experience. Washington, D.C.

Overseas Development Administration. 1989. Manual of Environmental Appraisal. London, United Kingdom.

Organisation for Economic Co-Operation and Development. 1986. "Environmental Assessment and Development Assistance." Environment Monograph (4):103.

____. 1990. "Good Practices for Environmental Impact Assessment of Development Projects." Meeting of the Working Party on Development Assistance and Environment, May 10-11. (Restricted to Participants). Paris, France.

Ofori-Cudjoe, S. 1990. "Environmental Impact Assessment in Ghana. An Ex-Post Evaluation of the Volta Resettlement Scheme: The Case of the Kpong Hydro-Electric Project." Environmentalist 10:115-126.

O'Riordan, T., and W. R. D. Sewell, eds. 1981. Project Appraisal and Policy Review. Chichester, United Kingdom: John Wiley and Sons.

O'Riordan, T. 1989. "The Impact of Environmental Impact Assessment on Decision-Making." In Environmental Impact Assessment, edited by V. T. Covello. Heidelburg, Federal Republic of Germany: Springer.

Pearce, D. W., E. Barbier, and A. Markandya. 1990. Sustainable Development: Economics and Environment in the Third World. Aldershot, United Kingdom: Elgar.

Pendse, Y. D., R. V. Rao, and P. K. Sharma. 1989. "Environmental Impact Assessment Methodologies: Shortcomings and Appropriateness for Water Resources Projects in Developing Countries." International Journal of Water Resources Development 5(4):252-259.

Pimentel, D. 1989. "Agriculture and Ecotechnology." In Ecological Engineering: An Introduction to Ecotechnology, edited by W. J. Mitsch and S. E. Jorgensen. New York: John Wiley and Sons.

Porter, A. L. and others. 1980. A Guidebook for Technology Assessment and Impact Analysis. New York: North Holland.

Prieur, M. 1984. "Les Etudes d'Impact en Droit Francais." Zeitschrift fur Umweltpolitik 4:367-388.

Pryde, P. R. 1987. "The Soviet Approach to Environmental Impact Analysis." In Environmental Problems in the Soviet Union and Eastern Europe, edited by F. B. Singleton. Boulder, Colorado: Lynne Rienner Publishers.

Rau, J. G., and D. C. Wooten, eds. 1980. Environmental Impact Analysis Handbook. New York: McGraw-Hill.

Repetto, R., ed. 1985. The Global Possible Resources, Development and the New Century. New Haven, Connecticut: Yale University Press.

Roberts, R. D., and T. M. Roberts, eds. 1984. Planning and Ecology. London, United Kingdom: Chapman and Hall.

Ross, W. A. 1987. "Evaluating Environmental Impact Statements." Journal of Environmental Management 25(2):137-148.

Rossini, F. A., and A. L. Porter, eds. 1983. Integrated Impact Assessment. Boulder, Colorado: Westview Press.

Sadler, B., ed. 1987. Audit and Evaluation in Environmental Assessment and Management: Canadian and International Experience. 2 Volumes. Hull, Quebec: Environment Canada.

____. 1980. Public Participation in Environmental Decision Making: Strategies for Change. Edmonton, Alberta: Environment Council of Canada.

Sammy, G. K. 1982. "Environmental Impact in Developing Countries." Ph.D. Dissertation, University of Oklahoma. Norman, Oklahoma.

Schibuola, S., and P. H. Byer. 1991. "Use of Knowledge-Based Systems for the Review of Environmental Impact Assessments." Environmental Impact Assessment Review 11:11-27.

Shrader-Frechette, K. S. 1985. Science Policy, Ethics, and Economic Methodology: Some Problems of Technology Assessment and Environmental-Impact Analysis. Boston, Massachusetts: Riedel.

Sigal, L. L., and J. W. Webb. 1989. "The Programmatic Environmental Impact Statement: Its Purpose and Use." The Environmental Professional 11(1):14-17.

Sonntag, N. C. and others. 1987. Cumulative Effects Assessment: A Context for Further Research. Ottawa, Canada: Ministry of Services and Supply.

Strickland, G. T. 1984. Hunter's Tropical Medicine. 6th edition. Philadelphia, Pennsylvania: W. B. Sanders.

Suter, G. W. and others. 1987. "Treatment of Risk in Environmental Impact Assessment." Environmental Management 11:295-303.

Tchobanoglous, G., H. Theisen, and R. Eliassen. 1977. Solid Wastes: Engineering Principles and Management Issues. New York: McGraw-Hill.

Tharun, G., N. C. Thanh, and R. Bidwell, eds. 1983. Environmental Management for Developing Countries. 3 Volumes. Bangkok, Thailand: Asian Institute of Technology.

Tidsell, C. 1986. "Cost-Benefit Analysis, the Environment and Informational Constraints in LDCs." Journal of Development 11:63-81.

Tomlinson, P., ed. 1987. "Environmental Audits: Special Edition." Environmental Monitoring and Assessment 8(3):183-261.

United Nations. 1990. Environmental Assessment Procedures in the U.N. System. London, United Kingdom: Environmental Resources Limited.

United Nations Environment Programme. 1982. The Use of Environmental Impact Assessment for Development Project Planning in ASEAN Countries. Bangkok, Thailand: Regional Office for Asia and the Pacific.

United Nations Environment Programme and the World Health Organization. 1989. Assessment of Urban Air Quality. London, United Kingdom: Global Environment Monitoring System.

United Nations and the United Nations Asian and Pacific Development Institute. 1980. Environmental Impact Statements: A Test Model Presentation, comp. C. Suriyakumaran. Bangkok, Thailand.

United States Council on Environmental Quality [and] Fish and Wildlife Service. 1980. Biological Evaluation of Environmental Impacts. Report FWS/OBS-80/26. Washington, D.C.

United States Department of Energy. 1986. Digest of Environmental and Water Statistics No. 9. Washington, D.C.: General Printing Office.

Vlachos, E. 1990. "Assessing Long Range Cumulative Impacts." In Environmental Impact Assessment, edited by V. T. Covello. Heidelburg, Federal Republic of Germany: Springer.

Wandesforde-Smith, G., and I. Moreira. 1985. "Subnational Government and Environmental Impact Assessment in the Developing World: Bureaucratic Strategy and Political Change in Rio de Janeiro." Brazilian Environmental Impact Assessment Review 5:223-238.

Ward, D. V. 1978. Biological Environmental Impact Studies: Their Theory and Methods. New York: Academic Press.

Warner, M. L., and E. H. Preston. 1974. Review of Environmental Impact Assessment Methodologies. Washington D.C.: United States Environmental Protection Agency.

Wathern, P. 1984. "Methods for Assessing Indirect Impacts." In Perspectives on Environmental Impact Assessment, edited by B. D. Clark and others. Dordrecht, The Netherlands: Riedel.

Wathern P. and others. 1987. "Assessing the Environmental Impacts of Policy: A Generalized Framework for Appraisal." Landscape and Urban Planning 14:321-330.

Wathern, P., ed. 1988. Environmental Impact Assessment: Theory and Practice. London, United Kingdom: Unwin.

Wenger, R. B., W. Huadong, and Ma Xiaoying. 1990. "Environmental Impact Assessment in the People's Republic of China." Environmental Management 14:429-439.

Westman, W. E. 1985. Ecology, Impact Assessment and Environmental Planning. New York: John Wiley and Sons.

Wetsone, G. S., and A. Rosencranz. 1984. Acid Rain in Europe and North America. Washington, D.C.: Environmental Law Institute.

Wilson, D. G., ed. 1977. Handbook of Solid Waste Management. New York: Van Nostrand Reinhold Company.

Wood, C. M., and V. Cazidellis. 1985. A Guide to Training Materials for Environmental Impact Assessment. Department of Town and Country Planning Paper 14. Manchester, United Kingdom: University of Manchester.

World Bank. 1991. Country Capacity to Conduct Environmental Assessments in Sub-Saharan Africa. Technical Department, Africa Region. Environmental Division Working Paper 1. World Bank, Washington, D.C.

___. 1990. Environmental Health Components for Water Supply, Sanitation and Urban Projects. World Bank Technical Paper 121. Washington, D.C.: World Bank.

World Health Organization. 1974. Health Project Management: A Manual of Procedures for Formulating and Implementing Health Projects. Geneva, Switzerland.

Yates, E. D. 1989. Environmental Impact Assessment: What it is and Why International Development Organizations Need it. Washington, D.C.: Council on Environmental Quality.

Young, K., ed. 1988. Women and Economic Development: Local, Regional and National Planning Strategies. New York: Berg (for Oxford and United Nations Educational, Scientific and Cultural Organization).

Ziyun, F. 1989. "Environmental Impact Assessment of Yangtze Valley Projects." International Water Power and Dam Construction 41:36-39.

Distributors of World Bank Publications

ARGENTINA
Carlos Hirsch, SRL
Galeria Guemes
Florida 165, 4th Floor-Ofc. 453/465
1333 Buenos Aires

AUSTRALIA, PAPUA NEW GUINEA, FIJI, SOLOMON ISLANDS, VANUATU, AND WESTERN SAMOA
D.A. Books & Journals
648 Whitehorse Road
Mitcham 3132
Victoria

AUSTRIA
Gerold and Co.
Graben 31
A-1011 Wien

BAHRAIN
Bahrain Research and Consultancy
Associates Ltd.
P.O. Box 22103
Manama Town 317

BANGLADESH
Micro Industries Development
Assistance Society (MIDAS)
House 5, Road 16
Dhanmondi R/Area
Dhaka 1209

Branch offices:
Main Road
Maijdee Court
Noakhali - 3800

76, K.D.A. Avenue
Kulna

BELGIUM
Jean De Lannoy
Av. du Roi 202
1060 Brussels

CANADA
Le Diffuseur
C.P. 85, 1501B rue Ampère
Boucherville, Québec
J4B 5E6

CHINA
China Financial & Economic
Publishing House
8, Da Fo Si Dong Jie
Beijing

COLOMBIA
Infoenlace Ltda.
Apartado Aereo 34270
Bogota D.E.

COTE D'IVOIRE
Centre d'Edition et de Diffusion
Africaines (CEDA)
04 B.P. 541
Abidjan 04 Plateau

CYPRUS
MEMRB Information Services
P.O. Box 2098
Nicosia

DENMARK
SamfundsLitteratur
Rosenoerns Allé 11
DK-1970 Frederiksberg C

DOMINICAN REPUBLIC
Editora Taller, C. por A.
Restauración e Isabel la Católica 309
Apartado Postal 2190
Santo Domingo

EL SALVADOR
Fusades
Avenida Manuel Enrique Araujo #3530
Edificio SISA, 1er. Piso
San Salvador

EGYPT, ARAB REPUBLIC OF
Al Ahram
Al Galaa Street
Cairo

The Middle East Observer
8 Chawarbi Street
Cairo

FINLAND
Akateeminen Kirjakauppa
P.O. Box 128
SF-00101
Helsinki 10

FRANCE
World Bank Publications
66, avenue d'Iéna
75116 Paris

GERMANY
UNO-Verlag
Poppelsdorfer Allee 55
D-5300 Bonn 1

GREECE
KEME
24, Ippodamou Street Platia Plastiras
Athens-11635

GUATEMALA
Librerias Piedra Santa
5a. Calle 7-55
Zona 1
Guatemala City

HONG KONG, MACAO
Asia 2000 Ltd.
6 Fl., 146 Prince Edward
Road, W.
Kowloon
Hong Kong

INDIA
Allied Publishers Private Ltd.
751 Mount Road
Madras - 600 002

Branch offices:
15 J.N. Heredia Marg
Ballard Estate
Bombay - 400 038

13/14 Asaf Ali Road
New Delhi - 110 002

17 Chittaranjan Avenue
Calcutta - 700 072

Jayadeva Hostel Building
5th Main Road Gandhinagar
Bangalore - 560 009

3-5-1129 Kachiguda Cross Road
Hyderabad - 500 027

Prarthana Flats, 2nd Floor
Near Thakore Baug, Navrangpura
Ahmedabad - 380 009

Patiala House
16-A Ashok Marg
Lucknow - 226 001

INDONESIA
Pt. Indira Limited
Jl. Sam Ratulangi 37
P.O. Box 181
Jakarta Pusat

ITALY
Licosa Commissionaria Sansoni SPA
Via Benedetto Fortini, 120/10
Casella Postale 552
50125 Florence

JAPAN
Eastern Book Service
37-3, Hongo 3-Chome, Bunkyo-ku 113
Tokyo

KENYA
Africa Book Service (E.A.) Ltd.
P.O. Box 45245
Nairobi

KOREA, REPUBLIC OF
Pan Korea Book Corporation
P.O. Box 101, Kwangwhamun
Seoul

KUWAIT
MEMRB Information Services
P.O. Box 5465

MALAYSIA
University of Malaya Cooperative
Bookshop, Limited
P.O. Box 1127, Jalan Pantai Baru
Kuala Lumpur

MEXICO
INFOTEC
Apartado Postal 22-860
14060 Tlalpan, Mexico D.F.

MOROCCO
Société d'Etudes Marketing Marocaine
12 rue Mozart, Bd. d'Anfa
Casablanca

NETHERLANDS
InOr-Publikaties b.v.
P.O. Box 14
7240 BA Lochem

NEW ZEALAND
Hills Library and Information Service
Private Bag
New Market
Auckland

NIGERIA
University Press Limited
Three Crowns Building Jericho
Private Mail Bag 5095
Ibadan

NORWAY
Narvesen Information Center
Book Department
P.O. Box 6125 Etterstad
N-0602 Oslo 6

OMAN
MEMRB Information Services
P.O. Box 1613, Seeb Airport
Muscat

PAKISTAN
Mirza Book Agency
65, Shahrah-e-Quaid-e-Azam
P.O. Box No. 729
Lahore 3

PERU
Editorial Desarrollo SA
Apartado 3824
Lima

PHILIPPINES
International Book Center
Fifth Floor, Filipinas Life Building
Ayala Avenue, Makati
Metro Manila

POLAND
ORPAN
Palac Kultury i Nauki
00-901 Warszawa

PORTUGAL
Livraria Portugal
Rua Do Carmo 70-74
1200 Lisbon

SAUDI ARABIA, QATAR
Jarir Book Store
P.O. Box 3196
Riyadh 11471

MEMRB Information Services
Branch offices:
Al Alsa Street
Al Dahna Center
First Floor
P.O. Box 7188
Riyadh

Haji Abdullah Alireza Building
King Khaled Street
P.O. Box 3969
Damman

33, Mohammed Hassan Awad Street
P.O. Box 5978
Jeddah

SINGAPORE, TAIWAN, MYANMAR, BRUNEI
Information Publications
Private, Ltd.
02-06 1st Fl., Pei-Fu Industrial
Bldg.
24 New Industrial Road
Singapore 1953

SOUTH AFRICA, BOTSWANA
For single titles:
Oxford University Press
Southern Africa
P.O. Box 1141
Cape Town 8000

For subscription orders:
International Subscription Service
P.O. Box 41095
Craighall
Johannesburg 2024

SPAIN
Mundi-Prensa Libros, S.A.
Castello 37
28001 Madrid

Librería Internacional AEDOS
Consell de Cent, 391
08009 Barcelona

SRI LANKA AND THE MALDIVES
Lake House Bookshop
P.O. Box 244
100, Sir Chittampalam A.
Gardiner Mawatha
Colombo 2

SWEDEN
For single titles:
Fritzes Fackboksforetaget
Regeringsgatan 12, Box 16356
S-103 27 Stockholm

For subscription orders:
Wennergren-Williams AB
Box 30004
S-104 25 Stockholm

SWITZERLAND
For single titles:
Librairie Payot
6, rue Grenus
Case postale 381
CH 1211 Geneva 11

For subscription orders:
Librairie Payot
Service des Abonnements
Case postale 3312
CH 1002 Lausanne

TANZANIA
Oxford University Press
P.O. Box 5299
Dar es Salaam

THAILAND
Central Department Store
306 Silom Road
Bangkok

TRINIDAD & TOBAGO, ANTIGUA BARBUDA, BARBADOS, DOMINICA, GRENADA, GUYANA, JAMAICA, MONTSERRAT, ST. KITTS & NEVIS, ST. LUCIA, ST. VINCENT & GRENADINES
Systematics Studies Unit
#9 Watts Street
Curepe
Trinidad, West Indies

UNITED ARAB EMIRATES
MEMRB Gulf Co.
P.O. Box 6097
Sharjah

UNITED KINGDOM
Microinfo Ltd.
P.O. Box 3
Alton, Hampshire GU34 2PG
England

VENEZUELA
Libreria del Este
Aptdo. 60.337
Caracas 1060-A

YUGOSLAVIA
Jugoslovenska Knjiga
P.O. Box 36
Trg Republike
YU-11000 Belgrade